T0139938

Progress in IS

"PROGRESS in IS" encompasses the various areas of Information Systems in theory and practice, presenting cutting-edge advances in the field. It is aimed especially at researchers, doctoral students, and advanced practitioners. The series features both research monographs that make substantial contributions to our state of knowledge and handbooks and other edited volumes, in which a team of experts is organized by one or more leading authorities to write individual chapters on various aspects of the topic. "PROGRESS in IS" is edited by a global team of leading IS experts. The editorial board expressly welcomes new members to this group. Individual volumes in this series are supported by a minimum of two members of the editorial board, and a code of conduct mandatory for all members of the board ensures the quality and cutting-edge nature of the titles published under this series.

More information about this series at https://www.springer.com/bookseries/10440

Andreas Kamilaris · Volker Wohlgemuth ·
Kostas Karatzas · Ioannis N. Athanasiadis
Editors

Advances and New Trends in Environmental Informatics

Digital Twins for Sustainability

Editors
Andreas Kamilaris
Research Centre on Interactive Media
Smart Systems and Emerging Technologies (RISE)
Nicosia, Cyprus

Department of Computer Science
University of Twente
Enschede, The Netherlands

Kostas Karatzas
School of Mechanical Engineering
Aristotle University of Thessaloniki
Thessaloniki, Greece

Volker Wohlgemuth
School of Engineering—Technology and Life
HTW Berlin—University of Applied Sciences
Berlin, Germany

Ioannis N. Athanasiadis
Department of Environmental Sciences
Wageningen University & Research
Wageningen, The Netherlands

ISSN 2196-8705 ISSN 2196-8713 (electronic)
Progress in IS
ISBN 978-3-030-61971-8 ISBN 978-3-030-61969-5 (eBook)
https://doi.org/10.1007/978-3-030-61969-5

This Springer imprint is published by the registered company Springer Nature Switzerland AG
The registered company address is: Gewerbestrasse 11, 6330 Cham, Switzerland

Preface

This book presents the main research results of the 34th edition of the long-standing and established international and interdisciplinary conference series on environmental information and communication technologies (EnviroInfo 2020).

The conference was held from 23 to 24 September 2020 virtually. It was organized by the Research Centre on Interactive Media, Smart Systems and Emerging Technologies (RISE), Nicosia, Cyprus, under the patronage of the Technical Committee on Environmental Informatics of the Gesellschaft für Informatik e.V. (German Informatics Society—GI). RISE is a research centre of excellence in Cyprus, aiming to empower knowledge and technology transfer in the region of South-East Mediterranean. It is a joint venture between the three public universities of Cyprus (University of Cyprus, Cyprus University of Technology and Open University of Cyprus), the Municipality of Nicosia and two renowned international partners, the Max Planck Institute for Informatics, Germany, and the University College London, UK.

This book presents a selection of peer-reviewed research papers that describe innovative scientific approaches and ongoing research in environmental informatics and the emerging field of environmental sustainability. Combining and shaping national and international activities in the field of applied informatics and environmental informatics, the EnviroInfo conference series aims at presenting and discussing the latest state-of-the-art development on information and communication technology (ICT) and environmental-related fields. A special focus of the conference was on digital twins and, in particular, the emerging research concept of digital twins for sustainability, where natural systems are twinned with digital replicas, to improve our understanding of complex socio-environmental systems through advanced intelligence. Sustainable digital twins of smart environments are also a flagship project of RISE.

The respective articles cover a broad range of scientific aspects including advances in core environmental informatics-related technologies, such as earth observation, environmental monitoring and modelling, big data and machine learning, robotics, smart agriculture and food solutions, renewable energy-based solutions, optimization of infrastructures, sustainable industrial/production processes and citizen science, as well as applications of ICT solutions intended to support societal transformation processes towards the more sustainable management of resource use, transportation and energy supplies.

We would like to thank all contributors for their submissions. Special thanks also go to the members of the programme and organizing committees, for reviewing all submissions. In particular, we like to thank our local organizers at RISE who responded fast and generated a digital twin of the physical conference and hosted it online. We also deeply appreciate the help and support of the Environmental Informatics Community that backed up our efforts to cope with the COVID-19 pandemic and to have a stimulating and productive online event. Last but not least, a warm thank you to our sponsors that supported the conference.

Finally, we wish to thank Mrs. Barbara Bethke and Mr. Christian Rauscher from Springer and the entire Springer production team for their assistance and guidance in successfully producing this book.

Nicosia, Cyprus
Berlin, Germany
Thessaloniki, Greece
Wageningen, The Netherlands
September 2020

Andreas Kamilaris
Volker Wohlgemuth
Kostas Karatzas
Ioannis N. Athanasiadis

Contents

Environmental Modelling, Monitoring and Information Systems

Urban Environments and Systems

Physical Environments

Industrial Environments and Processes

Designing for Sustainability: Lessons Learned from Four Industrial Projects

Patricia Lago, Roberto Verdecchia, Nelly Condori-Fernandez, Eko Rahmadian, Janina Sturm, Thijmen van Nijnanten, Rex Bosma, Christophe Debuysscher, and Paulo Ricardo

Abstract Scientific research addressing the relation between software and sustainability is slowly maturing in two focus areas, related to 'sustainable software' and 'software for sustainability'. The first is better understood and may include research foci like energy efficient software and software maintainability. It most-frequently covers 'technical' concerns. The second, 'software for sustainability', is much broader in both scope and potential impact, as it entails how software can contribute to sustainability goals in any sector or application domain. Next to the technical concerns, it may also cover economic, social, and environmental sustainability. Differently from researchers, practitioners are often not aware or well-trained in all four types of software sustainability concerns. To address this need, in previous work we have defined the Sustainability-Quality Assessment Framework (SAF) and assessed its viability via the analysis of a series of software projects. Nevertheless, it was never used by practitioners themselves, hence triggering the question: *What can we learn from the use of SAF in practice?* To answer this question, we report the results of practitioners applying the SAF to four industrial cases. The results

---

Disclaimer: the view expressed by the authors affiliated with the European Patent Office (EPO) is not necessarily that of the EPO.

P. Lago (✉) · R. Verdecchia · N. Condori-Fernandez
Vrije Universiteit Amsterdam, Amsterdam, The Netherlands
e-mail: p.lago@vu.nl

P. Lago
Chalmers University of Technology, Gothenburg, Sweden

E. Rahmadian
University of Groningen, Groningen, The Netherlands

J. Sturm
German Development Institute, Bonn, Germany

T. van Nijnanten
Vandebron, Amsterdam, The Netherlands

R. Bosma · C. Debuysscher · P. Ricardo
European Patent Office, Munich, Germany

A. Kamilaris et al. (eds.), *Advances and New Trends in Environmental Informatics*, Progress in IS, https://doi.org/10.1007/978-3-030-61969-5_1

show that the SAF helps practitioners in (1) creating a sustainability mindset in their practices, (2) uncovering the relevant sustainability-quality concerns for the software project at hand, and (3) reasoning about the inter-dependencies and trade-offs of such concerns as well as the related short- and long-term implications. Next to improvements for the SAF, the main lesson for us as researchers is the missing explicit link between the SAF and the (technical) architecture design.

Keywords Decision maps · Sustainability-quality model · Design concerns · Lessons learned · Industrial projects

1 Introduction

With the ever growing pervasiveness of software-intensive systems, numerous questions arose on how to effectively and efficiently develop and maintain a software system. This led to the establishment of a vast corpus of knowledge on how to design, implement, and evaluate software-intensive systems. Still, to date, most efforts in software engineering focus on the optimization of technical aspects of software systems. Nevertheless, recently new researches emerged questioning: *what makes a software-intensive system sustainable? and what makes it contribute to sustainability?*

The growing research interest in the topics related to software sustainability led to the definition of *sustainability-awareness* as a software quality requirement. Such concept blossomed from the joint effort of academic researchers to define what it means for a software-intensive system to be sustainable e.g. [4, 9, 18, 22], and what role software engineering plays in its establishment [5, 6, 17, 20].

Following the definition of Lago et al. [18] and Venters et al. [22], software sustainability can be characterized in distinct yet interdependent dimensions. Based on such concept of sustainability dimension, Condori et al. [9] refined the definition of four core dimensions of software sustainability, namely the economic, technical, social, and environmental ones.

The four dimensions of sustainability are included in the Sustainability Assessment Framework (SAF) [10], a framework and accompanying toolkit[1] proposed to support data-driven reasoning and evaluation of the different sustainability dimensions which characterize a software-intensive system. The SAF is composed of three main components: the Sustainability-Quality (SQ) Model [9], the architectural decision maps [16], and the related suite of metrics [8].

Addressing sustainability in software engineering has a very broad scope, as illustrated by two manifestos [1, 12]. SQ assessment, in particular, is spanning from the assessment of software energy efficiency (e.g. [21, 23]) to the evaluation of the maturity of whole organizations with respect to Green ICT (e.g. [13, 14]). In this work we focus specifically on supporting software architects and design decision

[1] SAF Toolkit, or Toolkit for short.

makers in the definition of sound SQ assessment. In this context, related works are relatively limited.

With a special focus on requirements engineering, Becker et al. [2] add '*individual*' as a fifth sustainability dimension in addition to the four sustainability dimensions used in this work. However, we argue that the social and individual dimensions share the same *social nature*. Furthermore, the first takes a broader perspective (e.g. organizations, society, stakeholder types), which is especially relevant in software architecture because it aims at capturing "the big picture". Considering the individual as an additional dimension is only appropriate when their concerns must be addressed. (e.g. in requirements engineering or human-computer interaction). Duboc et al. [11], in turn, define a framework for raising sustainability awareness and perform an evaluation that shows its effectiveness. This work can be seen as complementary to ours, by adding awareness creation as a first step followed by design decision making.

In order to create awareness on sustainability-quality requirements, we have conducted several empirical studies carried out with real-life projects in software companies [7, 10], but these studies focused only on the SQ model.

Finally, some works surveyed quality models and touched upon their relation to sustainability (e.g. [19]). Although still work in progress, to the best of our knowledge, our SAF Toolkit is the only providing concrete guidance for SQ design and assessment. Following previous validation of the SAF Toolkit [10], in this study we assess the experience of practitioners in applying it, with the dual goal of gathering their lessons learned as well as lessons to improve the Toolkit.

2 Background

The SAF was proposed to guide decision making from a software architect perspective. Of its components, the **Decision Maps (DMs)** essentially frame the *expected impact* of a software architecture on the relevant *sustainability concerns*. According to Lago [16], there are three types of *expected impacts*: *(i) Immediate impacts* refer to immediately observable changes. These are addressed within the current software project and are expected to be directly traceable to the architecture entities. *(ii) Enabling impacts* arise from use over time. This includes the opportunity to consume more (or less) resources, but also shorten their useful life by obsolescence or substitution. *(iii) Systemic impacts* refer to persistent changes observable at the macro-level (e.g. behavioral change, economic structural change).

The types of *sustainability concerns* reflect the corresponding four sustainability dimensions: *(i) Technical dimension* addresses the long-term use of software-intensive systems and their appropriate evolution in an execution environment that continuously changes. *(ii) Economic dimension* focuses on preserving capital and economic value. *(iii) Social dimension* focuses on supporting current and future generations to have the same or greater access to social resources by pursuing generational equity. For software-intensive systems, this dimension encompasses the

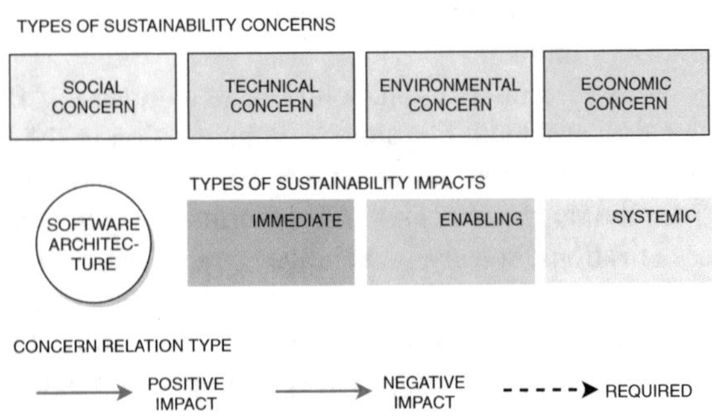

Fig. 1 Legend of the DM visual notation

direct support of social communities, as well as the support of activities or processes that indirectly create benefits for such communities. *(iv) Environmental dimension* aims at improving human welfare while protecting natural resources. For software-intensive systems, this dimension aims at addressing ecologic concerns, including energy efficiency and ecologic awareness creation. The relationships among design concerns are defined as *Effects*. We have three types of effects: *positive*, *negative*, and *required*. A legend of the visual notation used in the DMs is shown in Fig. 1.

Based on the ISO/IEC 25010 Standard [15], the **SQ Model** is made of a set of *Quality Attributes (QAs)* classified in the four sustainability dimensions, e.g. security in the technical dimension, energy efficiency in the environmental one. The QAs can be dependent. Such *dependency* can be of two types: (i) *inter-dimensional*, if it relates a pair of QAs defined simultaneously in two different dimensions (e.g. security defined in the *technical* dimension can influence security in the *social* dimension), and (ii) *intra-dimensional*, if a dependency exists between two different QAs defined within the same dimension (e.g. in the *technical* dimension, security may depend on reliability). Each QA of the SQ model is characterized by being measurable via a set of metrics.

The SQ model, as an Toolkit instrument, provides support to identify (i) sustainability concerns, and particularly those related to QAs; and (ii) the types of effect by means of the dependencies among QAs. In order to facilitate the creation of a DM, the following Toolkit instruments were created:

- A list of QAs in the SQ Model with their corresponding definitions and contributions to one or more sustainability dimensions [9].
- A set of dependency matrixes, representing the inter-dimensional dependencies in the SQ model.
- A decision graph, facilitating the correct identification of the types of impact.
- A custom library for the Draw.io editor tool,[2] used to draw the DMs (see Fig. 3) according to the visual notation in Fig 1.

[2]https://www.draw.io.

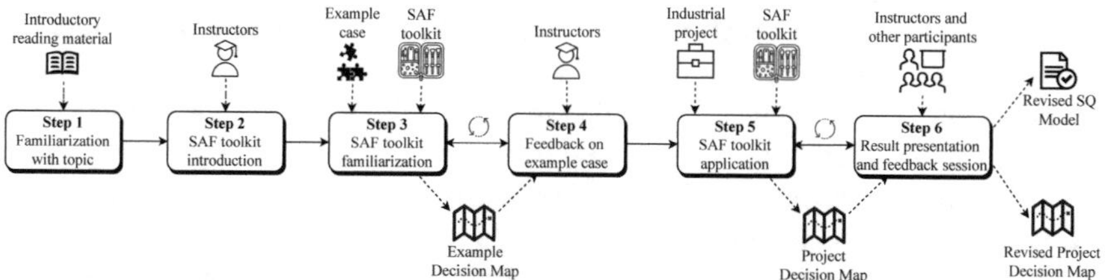

Fig. 2 Overview of the study design and execution

3 Study Design and Execution

In this section we document the design of our study and the details of the study execution. The focus of our study is to apply the SAF to concrete software innovation projects, with the goal of gathering lessons learned from both practitioner and researcher viewpoints. To do so, we carried out a set of working sessions, taking place over a week during the first graduate winter school "*Software and Sustainability: Towards an ethical digital society*"[3] at the Vrije Universiteit Amsterdam. In total, 6 participants were involved in the study, all with a consolidated industrial experience, ranging from 6 to 31 years, in sectors related to ICT and sustainability. The participants were involved in all steps of our study reported below, which constitute the outline of our study design. To gather data on the application of the SAF, we conducted educational sessions to provide participants with a sound understanding of the SAF and related concepts. Subsequently, participants applied the framework to concrete software innovation projects they were currently involved in. More in detail, the design of our study can be decomposed in 6 distinct steps, namely (i) preliminary familiarization with the topic of sustainability, (ii) introduction to the SAF toolkit, (iii) familiarization with the toolkit via a predefined hands-on example case, (iv) feedback on the example case execution. An eagle-eye overview of the process followed for our study is reported in Fig. 2, while the single steps composing the process are described in detail in the following.

Step 1: Familiarization with the topic. In this preliminary phase, participants were invited to study a small set of introductory material (i.e. [9, 16]), in order to get accustomed with the topics of ICT sustainability and the SAF prior to the first session.

Step 2: SAF Toolkit Introduction. In order to ensure that all participants possessed a sufficient level of background knowledge prior to the application of the SAF toolkit, two frontal lectures on the topic of sustainability and the SAF were conducted. Specifically, the first lecture focused on providing a sufficient level of knowledge on the notion of sustainability in software-intensive systems, with particular emphasis on the different dimensions of software-sustainability [18], how the dimensions can vary across different systems, and how the dimensions can

[3] https://tinyurl.com/yxemrk6c.

impact positively and negatively our society. The second lecture instead focused on the introduction of the SAF and related concepts, with the goal of providing participants with sufficient knowledge to concretely apply the SAF toolkit. The lectures lasted a total of 3.5 hours, and were carried out in an interactive fashion, i.e., by actively engaging participants in discussions via questions and requests for feedback. The involvement of participants in discussions allowed to ensure, in a lightweight and informal fashion, their assimilation of the presented concepts.

Step 3: SAF Toolkit Familiarization. After the establishment of a common background knowledge on the SAF, a preliminary phase of familiarization with the framework was carried out. In this step, the participants analyzed via the SAF toolkit an example case, detailing a software-intensive system implementing a school enrollment management process. Specifically, participants were divided into groups, in order to let them jointly work on the example case. This let participants independently apply their newly acquired knowledge of the SAF for the first time. During this phase, instructors were only marginally involved, e.g. to clarify doubts on the application of the SAF toolkit. The minor intervention of instructors during this phase was purposely enforced, in order to let participants critically think about the SAF toolkit application. At the end of this step, each group was required to produce an example decision map, i.e., a decision map of the example case generated by applying the SAF toolkit.

Step 4: Feedback on Example Case. Subsequent to the generation of the examples decision maps, a feedback session involving both participants and instructors was carried out. During this phase, each group of participants presented the example decision map they worked on, and the results were jointly discussed with the instructors. Instructors provided feedback on the example decision maps, followed by further guidance on how to refine the application of the SAF. In order to ensure that participants fully assimilated the SAF analysis process and the details entailed by its application, Steps 3 and 4 were repeated two times. This constituted a feedback loop in which participants refined their skills over two days, by working on the same example project and perfecting their example decision maps according to the feedback of the instructors.

Step 5: SAF Toolkit Application. After the participants refined their skills by applying the SAF to the example case, they proceeded to analyze via the SAF toolkit a concrete industrial project. As introductory phase of this step, participants were asked to pitch, through a short presentation, a concrete industrial project they are working on. This provided participants with the possibility to carry out the SAF analysis on a project they were interested in and familiar with. Additionally, such project selection process allowed to collect real-life data on the practical application of the SAF to industrial projects. In total, four working groups were formed during this preliminary phase. Each group worked on a shared industrial project, as further discussed in Sect. 4. The output of this phase consisted of a preliminary decision map for each industrial project considered. In the eventuality that participants felt the need to carry out adjustments of the SQ model to better fit their project, they were instructed to note down their modification, in order to discuss them in Step 6.

Step 6: Results Presentation and Feedback Session. Similar to the application of the SAF toolkit to the example case, its application to the industrial projects was characterized by a feedback loop. Specifically, in order to refine the project decision maps created in Step 5, ad-hoc sessions were carried out. During such sessions participants presented their results, and got feedback on how to use the SAF toolkit, and correct/refine their decision maps. Both instructors and participants discussed each decision map, in order to make the result discussion a collective educational experience. Steps 5 and 6 were repeated two times, resulting in a revised project decision map per group (see output in Fig. 2). Additionally, adjustments that were done by the participants to the SQ Model during Step 5, were jointly discussed in the last feedback session, leading to the refinement of the SQ Model itself.

4 Projects and Related Results

This Section presents the four industrial projects with the lessons learned from practitioners from their DMs (Sects. 4.1, 4.2, 4.3 and 4.4) and the SQ Model (Sect. 4.5).

4.1 Project P1: Sustainable Tourism in Indonesia

Project Description The Indonesian Government wants to create an online platform to share and analyze data for transitioning toward a sustainable tourism. This should facilitate information exchange, monitoring and data-driven decision making for all relevant stakeholders (e.g. ministries, Statistics Indonesia, state energy companies, touristic organizations).

Currently, Indonesia is witnessing great economic growth thanks to tourism, but it lacks policies and regulations to ensure tourism' social and environmental sustainability. Data sharing among the key stakeholders is not supported or enforced; the government carries out time- and effort-consuming manual surveys to collect information; and understanding of the important issues is limited.

In this project, the *sustainability goal* is to identify the network of design concerns that help balancing economic growth and social/environmental sustainability of the tourism sector in Indonesia.

Project DM (Fig. 3a) The most important design concerns (required by this project to be successful) are interoperability, adaptability, and the definition of law and regulations that boost stakeholder engagement. By analyzing the network of dependencies captured in the DM, in short-term we expect these three concerns to lead to greater impacts on other aspects, for instance (technical) usefulness, (economic) efficiency and (social) accountability. In turn, for the long-term this will affect all of the sustainability dimension. Despite the social risk related to the use of big data by stakeholders, the accomplishment of this project will be strategic for the Indonesian

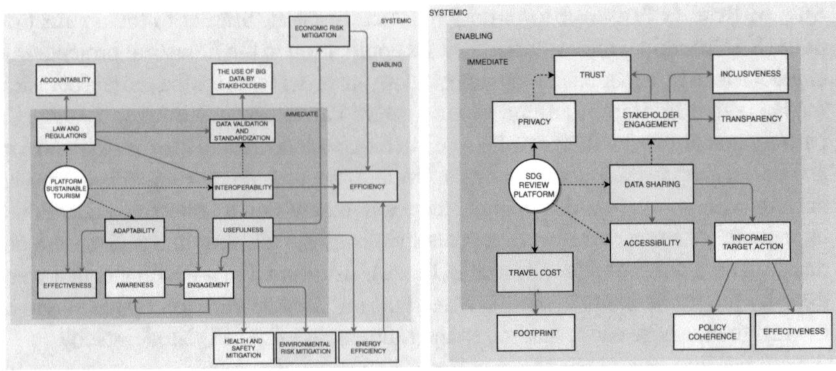

(a) DM of *"Sustainable Tourism in Indonesia"* (b) DM of *"SGD Review Platform"*

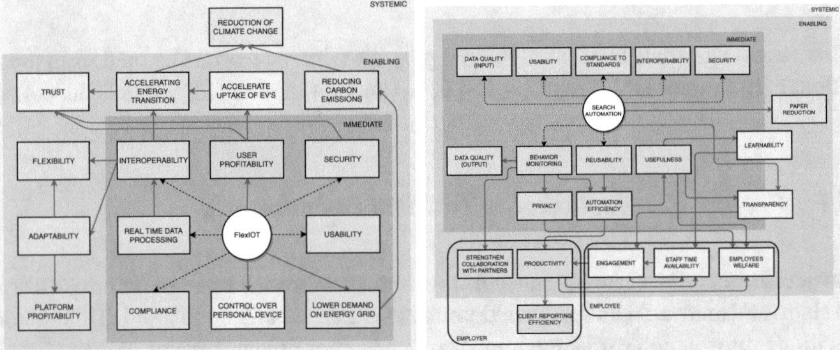

(c) DM of *"Energy Provisioning via Flexible IoT"* (d) DM of *"Search Automation for Literature Review"*

Fig. 3 Overview of the architectural DMs of the industrial projects

Government to contribute to achieving SDGs 6 (Clean Water), 7 (Clean Energy), 8 (Decent Work and Economic Growth), 11 (Sustainable Cities), 13 (Climate Change), and most importantly 17 (Partnership).

Lessons from the Practitioners

The Government is instrumental for engagement. The decision map shows that laws and regulations are necessary to trigger data sharing from the relevant stakeholders, which (thanks to interoperability and data standardization) can feed the platform with quality data automatically or semi-automatically. This will remove the need for manual surveys and the definition of guidelines for the various stakeholder.

4.2 *Project P2: SGD Review Platform*

Project Description This project explores the potential effects of creating an online platform for the United Nations (UN) to gather and share the progress of the member states with respect to the global Sustainable Development Goals (SDGs, [3]).

For the review of the progress towards the achievement of the 17 SDGs of the 2030 Agenda on Sustainable Development, an online platform could help to compare across sectors, countries and over time. Countries already present their voluntary national reports (VNRs) at the annual High-Level Political Forum on Sustainable Development (HLPF) in New York. Reviewing the progress towards sustainable development is necessary in order to see if countries are on track, if more or other measures are needed and where challenges continue to exist. A virtual platform could help to show a bigger picture over time and could allow different stakeholders to feed their data. The collection and quality of data is still challenging which hinders significant comparison and, in turn, the creation of effective actions and policies. In addition, factors hindering the creation of such online platform include the non-trivial need for global accessibility from all member states, and the creation of trustful data sharing on a global scale.

In this project, the *sustainability goal* is to identify the network of design concerns that help balancing the technical needs for global data sharing and accessibility, and the social concerns related to privacy, engagement, inclusiveness, and coherent and targeted policy-making.

Project DM (Fig. 3b) The design concerns are mainly of a social nature, such as the the engagement of different stakeholders (governments, non-governmental organisations) who can use the platform which leads to effects on inclusiveness and transparency. Other concerns refer to privacy issues that are touched through the collection and provision of data. There are technical concerns related to the accessibility of the platform to make sure data can be fed into the platform.

Lessons from the Practitioners

Social sustainability is crucial. The DM emphasizes that the number of potential positive effects on the social dimension is significantly higher than expected. This is of two types, the effect on privacy (being both positive thanks to ensuring it via the platform mechanisms, and negative due to the need to share data) and the need for engaging all relevant stakeholders.

UN meetings could be complemented by virtual meetings. A co-product of a successful online SDG Review Platform is that the meetings in New York could be complemented by virtual meetings and lead to more frequent exchanges between stakeholders. The extent to which the platform could trigger this change is uncertain. If this would happen, the effect of replacing physical- with virtual meetings is expected to have a significant lower footprint.

4.3 Project P3: Energy Provisioning via Flexible IoT

Project Description This project analyzes the sustainability concerns relevant for FlexIOT, a platform that manages the use of decentralized IoT-enabled assets (like the batteries of electric vehicles, heat pumps, or cooling systems) in order to balance supply and demand on the Dutch national high-voltage energy grid.

To accelerate the energy transition and reduce carbon emissions, we must replace fossil fuel plants with alternative solutions that use renewable resources and are equally reliable. Access to IoT-enabled assets owned by prosumers, however, requires their trust, profitability for all parties involved, and flexibility in terms of both adaptability and scalability, among others.

In this project, the *sustainability goal* is to identify the network of design concerns that help balancing the technical flexibility of energy provisioning with IoT-enabled assets, and the need for inspiring trust and enabling behavioral change.

Project DM (Fig. 3c) Most of the design concerns of immediate impact are of a technical nature as the technical functionalities of FlexIOT are essential to its existence. However, the DM also highlights two user-related concerns which are required for the long-term sustainability of the platform: (i) gaining the trust of the prosumers (concern that depends on the extent to which other multiple concerns are addressed—see the incoming effect-arrows), and (ii) guaranteeing immediate profitability for the prosumer.

In this respect, the trade-off for the prosumer is a minimized loss of autonomy in the usage of his/her shared device. Psychologically, consumers experience the reduction of such autonomy as a big risk. Hence, experience shows that we must address two important concerns: (i) it is essential to mitigate this perceived risk by applying a financial reward (even if this is not enough to counteract the negative effect of a loss of autonomy); and (ii) the platform must create (social) trust by showing proof of the (technical) security it ensures. To this end, FlexIOT uses blockchain technology to handle all consumers data.

The ultimate systemic goal of this project is to reduce climate change by applying a reliable alternative to fossil fuel plants. This can be effectuated by FlexIOT, which is able to control an exponential number of assets by adding different asset-types. Making flexible assets profitable for prosumers, reduces the barrier to purchase an IOT-enabled device, resulting in a faster uptake of these assets, and finally an acceleration of the reduction of carbon emissions.

Lessons from the Practitioners

Trust is a decisive concern. The longevity of the system relies on its log term profitability for the operator. This can be achieved through interoperability, which effects both the flexibility of the system, and it's adaptability. Meaning the system is able to both control different assets, as well as accept input from different systems in order to control these assets.

Interoperability is an essential growth enabler. Interoperability is positively affecting both adaptability and flexibility (the latter further enforced via adaptability). This shows that this technical requirement of the system is essential, as adaptability is crucial for the scalability of the platform, and hence its economic profitability.

4.4 Project P4: Search Automation for Literature Reviews

Project Description This project explores the socio-economic sustainability of a tool supporting researchers in performing literature reviews for external clients. The tool relies on machine learning algorithms trained on logged data about the manual search behavior of researchers. In the short term, it can provide them assistance by suggesting highly relevant papers or search terms; in the long term, it can (partially) automatize searches in suitable technical fields or for specific review studies, hence potentially resulting in higher work efficiency and faster reporting to clients.

Monitoring the search behavior of researchers, however, is a prerequisite to train the tool. As the logs show precisely how researchers work, without proper data protection it could be potentially misused by the research institutes they work for, e.g. to pinpoint less efficient employees or enforce higher productivity targets.

In this project, the *sustainability goal* is to identify the network of design concerns that help balancing the economic interests of the employer (the research institute) and the employees' social concerns ensuring workplace wellbeing.

Project DM (Fig. 3d) As we can observe in Fig. 3, the vast majority of immediate impact design concerns are of a technical nature including, among others, aspects related to data quality, automation efficiency, interoperability, and security. This highlights the core technical nature of this project. Interestingly, the DM clearly outlines that these technical concerns are expected to have an influence on other concerns belonging to different sustainability dimensions, of enabling and systematic impact.

Specifically, social concerns are characterized by an enabling impact, and are derived for the most part directly from the system's technical concerns. Such social concerns are related with the employees. Differently, also two economic concerns have an enabling impact, but this time the concerns are associated to the employer rather than the employee.

From the DM we can also observe how the project has a direct positive effect on paper reduction in the environmental sustainability dimension.

Finally, we can see that the only concern with a systemic impact is related to the economic dimension and reflects the end goal of the project.

Lessons from the Practitioners

Socio-economic concerns conflict between stakeholders. The DM emphasizes that the sustainability of the envisaged tool depends on the balance between the employees' social concerns, and the economic concerns of the organization. Such a balance is crucial to engage the employees. Given that the organizational culture is of ensuring employee welfare by rewarding employees for their dedication, an option is to equally reinvest the efficiencies resulting from automation in staff enrichment activities and economic productivity.

4.5 *Revised Sustainability Quality Model*

During the design of the DMs in the projects, participants could refine the SQ Model by creating ad hoc definitions of QAs specific to a sustainability dimension. In other words, participants could use either the standard definition of the QA provided in the ISO/IEC 25010 standard [15], or re-define the QA according to the specific sustainability dimension and the context of their project. Additionally, while an initial mapping of QAs to sustainability dimensions was provided to the participants (see coloured cells in Table 1), such mapping was not enforced, i.e., participants could add additional mappings according to their specific needs. In this section, we report the results of such process, which are schematically reported in Table 1.

As we can observe in Table 1, in 3 out of 4 projects the SQ Model was used to identify sustainability concerns and types of effects among related QAs. Participants of project P2 opted not to use the SQ model, as they deemed themselves not confident enough with the ISO/IEC 25010 standard to carry out the analysis. In total, 21 definitions of QAs were used, of which 14 by following the standard definition, and

Table 1 Sustainability-quality model analysis results (Colour = Mapping of QA to sustainability dimension, SP# = Standard definition of QA [15] used for project #, CP#-ID# = Custom definition ID# used for project #)

Characteristics	Attributes	Definition according to [6]	TECH	ENV	ECON	SOC
Compatibility	Interoperability	a system can exchange information with other systems and use the information that has been exchanged.	SP3			SP1
Context coverage	Flexibility	system can be used in contexts beyond those initially specified in the requirements.				CP3-1
Effectiveness	Effectiveness	accuracy and completeness with which users achieve specified goals.	CP4-1			SP1
Efficiency	Efficiency	resources expended in relation to the accuracy and completeness with which users achieve goals.				CP1-1
Freedom from risk	Economic risk mitigation	system mitigates the potential risk to financial status in the intended contexts of use.		SP3	CP1-2	
	Environmental risk mitigation	system mitigates the potential risk to property or the environment in the intended contexts of use.				SP1
	Health and safety risk mitigation	system mitigates the potential risk to people in the intended contexts of use.				SP1
Functional suitability	Functional appropriateness	the functions facilitate the accomplishment of specified tasks and objectives.	SP4			
	Functional correctness	system provides the correct results with the needed degree of precision.			SP4	
Maintainability	Modifiability	system can be effectively and efficiently modified without introducing defects or degrading existing product quality	SP4			
Performance efficiency	Time behaviour	response, processing times and throughput rates of a system, when performing its functions, meet requirements.	CP3-2, SP4			
Portability	Adaptability	system can effectively and efficiently be adapted for different or evolving hardware, software or usage environments.	SP3			
	Replaceability	product can be replaced by another specified software product for the same purpose in the same environment.				SP1
Satisfaction	Trust	stakeholders has confidence that a product or system will behave as intended.				CP3-3
	Usefulness	user is satisfied with their perceived achievement of pragmatic goals.				SP1
Security	Integrity	system prevents unauthorized access to, or modification of, computer programs or data.	SP3			
Usability	User error protection	system protects users against making errors.	CP3-4			

8 using ad hoc definitions. The most frequently considered dimensions of the SQ Model result to be the technical dimension (9/21) and social one (9/21). The depth in which the technical dimension is considered reflects the emphasis on technical concerns which characterizes projects P3 and P4. Similarly, the high recurrence of QAs mapped to the social dimension can be traced back to the relevance of the social dimension in SP1. Overall, QAs were only marginally mapped to the economic and environmental dimension. Interestingly, the environmental dimension was mapped to a single QA, which was not identified in the technical-action-research with which the SAF was validated [10]. The findings of this study will be further considered in order to refine the SQ Model, by considering the feedback of the participants, their results, and the context of the projects. For completeness, the ad hoc definitions provided by the participants is documented in Appendix 7.

5 Lessons to Improve the SAF Toolkit

The following summarizes the most important lessons we as researchers have learned throughout the whole week and which will help us improving the Toolkit. In particular, at the end of Step 6 we collected general feedback from the participants, as well as our own general observations from the way the participants worked at their project. Our main lessons learned are:

- Project P2 showed the need to use the "requires" relationship between concerns, too. This suggests that sustainability concerns may have a mix of inter-dependent effects (that can be part of a sustainability measure) and requirements (that should be satisfied by the implemented system, with no measure attached). While this does not require major changes in the DM notation, it plays an important role when concrete metrics are assigned to SQ measurements.
- Due to their unfamiliarity with the ISO/IEC 25010 standard, participants of Project P2 did not make use of the SQ model. This points to the need of a more in-depth training on the standard and related concepts, in order to ensure that all participants possess sufficient confidence to carry out the analysis via the SQ model.
- Project P4 extended the DM notation by clustering the concerns from the two stakeholders that are in conflict. This extension helped framing the presence of the conflicting stakes, and highlighting the chain of positive and negative effects that need balancing. In general, this shows that different perspectives can be illustrated also within a single view illustrated by a DM. Accordingly, we learn that different stakeholder perspectives can be captured both within a DM (when e.g., the network of concerns is simple enough), and with multiple DMs, one per stakeholder, when the complexity of the network of concerns hinders reasoning and decision making.
- In spite of the diversity in both the various projects and the expertise of the participants, a generalized surprising factor was that the DMs helped uncover the hidden social-sustainability concerns. The participants all agreed that social sustainability

is often left implicit while playing an instrumental role for achieving the target sustainability goals.

- In general, the participants all agreed that the Toolkit is a powerful instrument to (i) sharpen the design space, (ii) zoom out the details of the project at hand and gaining a broad perspective to spark *new* insights, (iii) facilitate informed choices, and (iv) communicate *what needs to be done* (including risks and benefits) with stakeholders with different concerns and expertise.
- The participants also agreed on a weakness of the Toolkit, namely the missing link between DMs and the (technical) architecture design views which are customary in software projects. We are happy to hear this as this is part of our ongoing and future research.

6 Conclusions and Future Work

This paper reports on a multi-case study where practitioners applied the SAF to four industrial cases. Our goal was to understand *What can we learn from the use of SAF in practice?* To this aim, we operationalized the SAF with the associated Toolkit instruments.

In spite of this being a single study with a relatively limited size (6 practitioners and 4 industrial cases), the results are very encouraging and suggest that the SAF can be readily used in practice, but that it needs further research (especially to define sound SQ metrics, and the explicit link between DMs and the (technical) software architecture elements and related views) to close the gap between design decision making and architecting.

The feedback we received from the practitioners indicates that the SAF helps them in (1) creating a sustainability mindset in their practices, (2) uncovering the relevant SQ concerns for the software project at hand, and (3) reasoning about the inter-dependencies and trade-offs of such concerns as well as the related short- and long-term implications. In addition, we could identify a number of lessons learned (described in Sect. 5) that will help us improving the SAF.

As future work we will continue training practitioners in using the SAF, with a dual benefit: they learn how to embed sustainability-quality in their software practices; we learn from them what needs to be included in the SAF Toolkit.

7 Appendix: Custom Quality Attributes Definitions

CP1-1 (Efficiency): *"Resources expended in relation to the accuracy, completeness and also less cost/time/human resources to conduct the research"*
CP1-2 (Economic Risk Mitigation): *"Mitigates risk to financial and economy for national/local level"*

CP3-1 (Flexibility): *"The system can be used in contexts beyond those initially specified in the requirements, such as controlling different assets"*
CP3-2 (Time Behaviour): *"Response, processing times and throughput rates of a system, when performing its functions, is real-time"*
CP3-3 (Trust): *"Users have confidence that a product or system will behave as intended."*
CP3-4 (User Error Protection): *"System protects users against making errors by being as intuitive as possible"*
CP4-1 (Effectiveness): *"Complies data quality requirements both in input and output"*
CP4-2 (Confidentiality): *"The system ensures that data are accessible only to those authorized to have access. Additionally, data should not be used for negative reporting, but only for improving efficiency."* Note: This QA was re-defined in Project P4 but not included in the corresponding DM.

References

1. Becker, C., Chitchyan, R., Duboc, L., Easterbrook, S., Penzenstadler, B., Seyff, N., Venters, C.: Sustainability design and software: the Karlskrona Manifesto. In: International Conference on Software Engineering: Software Engineering in Society (ICSE-SEIS), vol. 2, pp. 467–476 (2015)
2. Becker, C., Betz, S., Duboc, R.C.L., Easterbrook, S., Penzenstadler, B., Seyff, N., Venters, C.: Requirements: The key to sustainability. IEEE Software **33**(1) (2016)
3. Biermann, F., Kanie, N., Kim, R.E.: Global governance by goal-setting: the novel approach of the UN Sustainable Development Goals. Current Opin. Environ. Sustain. **26–27**, 26–31 (2017)
4. Calero, C., Moraga, M.A., Bertoa, M.F.: Towards a Software Product Sustainability Model. arXiv (2013). https://arxiv.org/abs/1309.1640
5. Chitchyan, R., Noppen, J., Groher, I.: What can software engineering do for sustainability: Case of software product lines. In: International Workshop on Product Line Approaches in Software Engineering. pp. 11–14 (2015)
6. Condori-Fernández, N., Lago, P.: The influence of green strategies design onto quality requirements prioritization. In: International Working Conference on Requirements Engineering: Foundation for Software Quality, pp. 189–205 (2018)
7. Condori-Fernández, N., Lago, P.: Towards a software sustainability-quality model: Insights from a multi-case study. In: International Conference on Research Challenges in Information Science (RCIS), pp. 1–11 (2019)
8. Condori-Fernández, N., Bagnato, A., Kern, E.: A focus group for operationalizing software sustainability with the MEASURE platform. In: International Workshop on Measurement and Metrics for Green and Sustainable Software Systems (MeGSuS) (2018)
9. Condori-Fernández, N., Lago, P.: Characterizing the contribution of quality requirements to software sustainability. J. Syst. Software **137**, 289–305 (2018)
10. Condori-Fernández, N., Lago, P., Luaces, M.R., Places, A.S., Folgueira, L.G.: Using Participatory Technical-action-research to validate a Software Sustainability Model. In: International Conference on ICT for Sustainability. ICT4S, CEUR-WS (2019)
11. Duboc, L., Betz, S., Penzenstadler, B., Kocak, S.A., Chitchyan, R., Leifler, O., Porras, J., Seyff, N., Venters, C.C.: Do we really know what we are building? raising awareness of potential sustainability effects of software systems in requirements engineering. In: International Requirements Engineering Conference (RE). IEEE, New York (2019)

12. Fonseca, A., Kazman, R., Lago, P.: A Manifesto for energy-aware software. IEEE Software **36**(6), 79–82 (2019)
13. Hankel, A., Heimeriks, G., Lago, P.: Green ICT adoption using a maturity model. Sustain.: Science Pract. Policy **11**(24), 7163 (2019)
14. Hankel, A., Oud, L., Saan, M., Lago, P.: A maturity model for green ICT: The case of the SURF green ICT maturity model. In: International Conference on Informatics for Environmental Protection (EnviroInfo), pp. 33–40. BIS Verlag (2014)
15. ISO/IEC: ISO/IEC 25010 - Systems and software engineering - System and software quality models. Tech. rep. (2010)
16. Lago, P.: Architecture design decision maps for software sustainability. In: IEEE/ACM International Conference on Software Engineering (ICSE), pp. 61–64 (2019)
17. Lago, P., Kazman, R., Meyer, N., Morisio, M., Müller, H.A., Paulisch, F.: Exploring initial challenges for green software engineering. SigSoft SEN **38**(1), 31–33 (2013)
18. Lago, P., Koçak, S.A., Crnkovic, I., Penzenstadler, B.: Framing sustainability as a property of software quality. Commun. ACM **58**(10), 70–78 (2015)
19. Zakaria, N.H. et al.: User Centric Software Quality Model For Sustainability: A Review. https://doi.org/10.18178/lnse.2016.4.3.250 (2016)
20. Penzenstadler, B.: Infusing green: requirements engineering for green in and through software systems. In: International Workshop on Requirements Engineering for Sustainable Systems (RE4SuSy), pp. 44–53 (2014)
21. Pihkola, H., Hongisto, M., Apilo, O., Lasanen, M.: Evaluating the energy consumption of mobile data transfer—from technology development to consumer behaviour and life cycle thinking. Sustainability **10**(7) (2018)
22. Venters, C., et. al: The blind men and the elephant: towards an empirical evaluation framework for software sustainability. J. Open Res. Software **2**(1) (2014)
23. Yan, M., Chan, C., Gygax, A., Yan, J., Campbell, L., Nirmalathas, A., Leckie, C.: Modeling the total energy consumption of mobile network services and applications. Energies **12**, 184 (2019)

Web Tool for the Identification of Industrial Symbioses in Industrial Parks

Anna Lütje, Sinéad Leber, Jonas Scholten, and Volker Wohlgemuth

Abstract Industrial Symbiosis (IS) is a systemic and collaborative business approach to optimize cycles of material and energy by connecting the supply and demand of various industries. IS provides approaches for advanced circular/cascading systems, in which the energy and material flows are prolonged for multiple utilization within industrial systems in order to increase resource productivity and efficiency. This study aims to present the conceptual IT-supported IS tool and its corresponding prototype, developed for the identification of IS opportunities in IPs. This IS tool serves as an IS facilitating platform, providing transparency among market players and proposing potential cooperation partners according to selectable criteria (e.g. geographical radius, material properties, material quality, purchase quantity, delivery period). So this IS tool builds the technology-enabled environment for the processes of first screening of IS possibilities and initiation for further complex business-driven negotiations and agreements for long-term IS business relationships. The central core of the web application is the analysis and modelling of material and energy flows, which refer to the entire industrial park as well as to individual companies. Methods of Material Flow Analysis (MFA) and Material Flow Cost Accounting (MFCA) are used to identify possible input–output- and supply–demand matchings. The second central core of the web application is the identification of existing and potential cooperation partners for the development of IS networks. In order to achieve this, a combinatorial approach of Social Network

---

A. Lütje (✉)
Institute of Environmental Communication, Leuphana University Lüneburg, Universitätsallee 1, 21335 Lüneburg, Germany
e-mail: anna.luetje@htw-berlin.de

A. Lütje · S. Leber · V. Wohlgemuth
Industrial Environmental Informatics Unit, Department Engineering—Technology and Life, Hochschule für Technik und Wirtschaft (HTW) Berlin, University of Applied Science, Treskowallee 8, 10318 Berlin, Germany

J. Scholten
Department Engineering—Energy and Information, Hochschule für Technik und Wirtschaft (HTW) Berlin, University of Applied Science, Computer Engineering, Treskowallee 8, 10318 Berlin, Germany

A. Kamilaris et al. (eds.), *Advances and New Trends in Environmental Informatics*, Progress in IS, https://doi.org/10.1007/978-3-030-61969-5_2

Analysis (SNA) and a deposited geographical map are inserted, so that same suppliers and recycling/disposal companies can be detected.

Keywords Industrial ecology · Industrial symbiosis · Material flow analysis · Material flow cost accounting · Social network analysis · Life cycle assessment

1 Introduction

Human activities have been major driving forces of global changes such as climate change, environmental pollution and increasing scarcity of resources, so that today´s era is called "the Anthropocene" [8]. In order to meet these challenges adequately, concepts such as Industrial Symbiosis (IS) are seen as a substantial key enabling factor for resource efficiency and circularity, contributing to the trajectory of sustainable development [14].

IS is covered by the scientific field of Industrial Ecology [5]. It is a systemic and collaborative business approach to optimize cycles of material and energy by connecting the supply and demand of various industries, while generating ecological, technical, social and economic benefits [4, 6, 10, 15, 19, 35, 39]. IS provides approaches for advanced circular/cascading systems, in which the energy and material flows are prolonged for multiple utilization within industrial systems in order to increase resource productivity and efficiency. The most cited definition is from Chertow [5], "IS engages traditionally separate industries in a collective approach to competitive advantage involving physical exchange of materials, energy, water and byproducts". According to Lombardi and Laybourn [24], "IS engages diverse organizations in a network to foster eco innovation and long-term culture change. Creating and sharing knowledge through the network yields mutually profitable transactions for novel sourcing of required inputs, value-added destinations for non-product outputs, and improved business and technical processes".

In the IS context, many companies approach their "waste" streams as new business opportunities or extended business models. For example, waste heat/exceeding steam can be forwarded to other companies, turning the originator company to an energy supplier [13, 31, 33, 42]. A smeltery in China recovered raw materials out of gaseous waste/aerosol, sludge/mud and solid waste [43]. The Guitang Group in China used their sludge as the calcium carbonate feedstock to a new cement plant, while reducing residual and waste flows, solving a disposal problem [44]. Gaseous waste streams such as fly ash can be used as cement additive [9, 12, 17] or soil additive [2, 32]. Waste water from a company that processes food such as olives, cereals, fruit and vegetables can be further used as fertilizer [7, 32], and for the irrigation of agricultural land, the respective organic residual (solid) waste can be further processed to animal/fish feeding (material utilization) [1, 3] or biogas and biofuel (energetic utilization) [1].

The aim of an IS is to establish virtually closed energy and material cycles/extended cascading systems through cooperation between companies. Industrial agglomerations and industrial parks (IPs) can be considered a favourable starting point to generate a first germ cell [6, 16, 20, 37, 41], as IS is predominantly based on collaboration and synergistic opportunities revealed by geographical proximity [6]. IS exhausted IPs mimic ecological dynamic systems concerning a development of a resilient system and an optimized use of resources by the cascading utilization of material and energy flows.

IS can occur spontaneously due to economic motivation, but just up to a certain point, then it needs to be further driven to exhaust its full potential [31]. This implies a systematized approach for cross-industry collaboration within a community through inter-organizational communication and information exchange [18, 22, 34]. The more complex the system to be considered, the more relevant computer-aided solutions become in order to map energetic and material movements or to facilitate cross-company collaborations. In order to identify and develop IS, instruments from the field of Information and Communication Technologies (ICT) play a central role. So the inter-company information flows for the identification of IS opportunities can be facilitated by an information platform [21, 36].

2 Method

This study aims to present the conceptual IT-supported IS tool and its corresponding prototype, developed for the identification of IS opportunities in IPs. The content concept of the IS tool has been designed based on previous research activities, concerning the development of an initial basic framework for the software/system development process phase of requirements engineering (RE), also called requirements analysis. RE is one of the main activities of the software or system development process, which defines the requirements for the system to be developed with the help of a systematic procedure from the project idea to the goals to a complete set of requirements. This was underpinned by extensive systematic literature research and analysis of existing case studies as well as interviews with IS experts and practitioners.

This study addresses the development of the IT concept and prototyping phase for an IT-supported IS tool for the identification of IS opportunities in IPs. Prototyping is an essential part of the software design process. A prototype is a first version of a software system that demonstrates concepts, tries out design options and learns more about upcoming problems and respective possible solutions. A prototype is particularly suitable for identifying changes that may be necessary or potential improvements at an early stage. During prototyping, both the graphical representation and the core functionalities of the system can be implemented. Prototypes provide the starting point for new requirements and can identify areas with strengths and weaknesses in the software. In addition, a prototype can be used to investigate the behavior of the system when several functionalities are combined (Fig. 1).

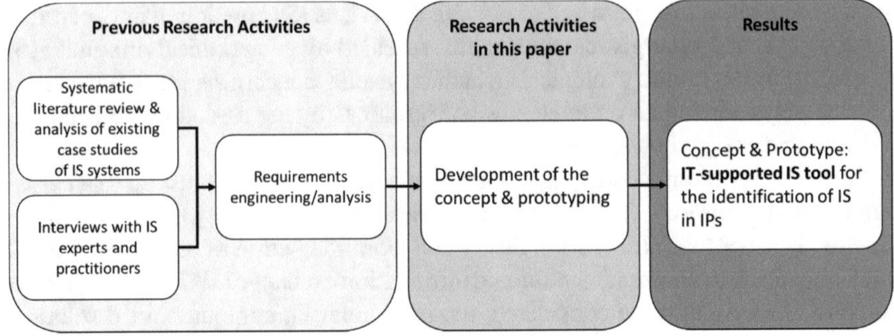

Fig. 1 Design of the research approach

3 Web Tool for the Identification of Industrial Symbioses Within an Industrial Park

3.1 Quantitative Methods for IS Identification

Various quantitative methods can be applied for the process of identifying IS opportunities. A previously conducted analysis of existing IS case studies revealed, that common methods such as emergy analysis,[1] Material Flow Analysis (MFA),[2] Material Flow Cost Accounting (MFCA),[3] Life Cycle Assessment (LCA)[4] and Social Network Analysis (SNA)[5] were used to investigate the current state system and to deduce possible IS activities [26]. For instance, Bain et al. [2] conducted a case study in an industrial area in South India, using MFA to analyze the recovery, reuse and recycling of industrial residuals and to identify existing symbiotic connections within this area. Applying MFA, Chertow [7] did a multiyear investigation of industrial sites in Puerto Rico between 2001 and 2007, in order to develop IS scenarios focused on utility sharing, joint service provision and by-product exchanges. LCA was used by

[1]Emergy is an expression of all the energy consumed in direct and indirect transformations in the processes to generate a product or service, therefore emergy analysis converts the thermodynamic basis of all forms of energy, resources and human services into equivalents of a single form of energy (usually solar emjoules).

[2]MFA quantifies the input and output flows and stocks of materials and energy for each process of the system under consideration in physical units (e.g. kg).

[3]MFCA quantifies the input and output flows and stocks of materials and energy in physical and monetary units, especially the material losses, non-/by-product and waste flows are associated an economic value (standardized to ISO 14051).

[4]LCA quantifies the input and output flows and stocks of materials and energy of entire product life cycles and assesses the associated environmental impacts, such as global warming and eutrophication potential (standardized to ISO 14040).

[5]SNA investigates social structures of networks and characterizes elements within the network in terms of nodes (e.g. individual actors, companies, people) and the connecting ties or links (relationships or interactions).

Sokka et al. [35] who studied a Finnish Forest Industry Complex around a pulp and paper mill to identify potential IS synergies. While Martin [30] explored IS in the biofuel industry in Sweden with LCA to detect IS opportunities and quantify the environmental performance. Ulhasanah and Goto [38] used a combinatorial approach of MFA, MFCA and LCA in a case study of cement production in Indonesia, in order to derive IS activities.

The research methods on IS systems have been extended from energy and material flows and its associated costs to social aspects, broadening the perspective to collaboration and the relationships among the IS entities, which can profoundly determine the effectiveness and efficiency of the entire IS system. SNA provides the investigation of the structure of IS systems and the (power) relationships of the entities involved. For instance, Song et al. [36] analyzed the Gujiao eco-industrial park in China and found out that SNA reveals IS potentials to develop more synergy linkages, identifying key/anchor actors in the network and their relation/context to exchanges of material, energy and (waste) water. Doménech and Davies [11] analyzed trust relationships in production or business networks in an empirical study of IS, using SNA.

3.2 *Prototype*

All previously collected information was incorporated into the design and prototypical development of the IT-supported IS tool. The prototypical web application was developed according to the REST[6] architecture. Technologies, frameworks and interfaces (APIs) from the field of web development were used such as AngularJS[7] (referring to the MVC architecture[8]), Spring Boot,[9] MySQL[10] and JPA.[11]

---

[6]REST stands for REpresentational State Transfer, enabling the realization of web services. Data is transferred via HTTP without the need for an additional transport layer such as SOAP (Simple Object Access Protocol) or session management via cookies. The clients send their requests to the server. The server processes them and returns a corresponding response. This communication is regarded in REST as a transfer of representations of resources. An application can interact with a resource if it knows the identifier of the resource and the action to be performed and can interpret the format of the returned information (representation). The server makes its external capabilities available not as services, but as resources that are identified via a URI (Uniform Resource Identifier).

[7]AngularJS is a library written in JavaScript for the development of dynamic web applications.

[8]The MVC architecture is a programming methodology with three core components: a model, a view, and a controller.

[9]Spring Boot is an open source framework for simplified application development with Java/Java EE.

[10]MySQL is a relational database management system. It is available as open source software as well as a commercial enterprise version for various operating systems and forms the basis for many dynamic websites.

[11]Java Persistence API (JPA) enables database access and object-relational mapping. This enables an object-oriented view of tables and relationships in a relational database management system (RDBMS). JPA can be used to work with objects instead of SQL statements.

In the front end, the service is designed as a single page application (SPA) to give it a look and feel of a desktop program. This is realised by using the Angular JS framework. It provides the developer with an MVC pattern, which makes it easy to extend, maintain and test the software.

For prototype development it is vital to have a running service as quickly as possible without lacking a proper design. Therefore, the choice was made to use the CSS framework Bootstrap. A responsive design can be implemented by making a few changes to the HTML elements.

Sankey.js is an extension of the Data Driven Documents (D3) library and is used to create the Sankey diagrams. The interactive image is generated by a JSON definition. With the D3 library as the core later implementations of new graphical representations can easily be integrated.

Leaflet is an open source library which enables the creation of interactive maps. By adding the ui-leaflet directive into Angular JS the map can be directly integrated with the <leaflet> HTML tag. It allows the display of tiled web maps which can be zoomed and dragged by mouse. As an overlay its features vary from clickable markers, CSS popups, circles and polygons.

The core of the backend is created by Spring Boot. The Java framework includes the implementation of a RESTful Webservice as well as dependency injection. Endpoints can be easily configured. The communication between client and server are made in JSON format. Data persistence is achieved by a MySQL database. By using the Java Persistance API (JPA) changes can be made by sending objects instead of SQL queries.

The web application serves as a tool for the analysis of material and energy flows as well as the detection of possible symbiosis partners within an industrial area. In this concept, three common methods were chosen (MFA, MFCA, SNA) due to simplified first IS screening reasons and user-friendly applicability of the IS tool. The implementation of other methods such as LCA or emergy analysis requires specific method knowledge and access to external (costly) environmental databases, that is why these kind of methods are not considered in this proposed IS tool.

The central core of the application is the analysis and modelling of material and energy flows, which refer to the entire industrial site as well as to individual companies. Methods of MFA and MFCA are used in the web application to identify possible input–output- and supply–demand matchings [25].

The second central core of the application is the identification of existing and potential cooperation partners for the development of IS networks. In order to achieve this, a combinatorial approach of SNA and a deposited geographical map are inserted, so that same suppliers and recycling/disposal companies can be detected, hence, the IS network can be intensified [27].

The web tool supports the functions of:

(1) Evaluating material and energy movements in companies and the entire IP (MFA)
(2) Evaluating the costs of material and energy flows in each company, especially waste streams are attributed an economic value (MFCA)

(3) Identification of material and energy related input–output and supply–demand matchings among the participating entities
(4) Visualizing the structure and relationships of the entities, embedded in an IP, and connected external suppliers or disposal/recycling companies (SNA)
(5) Identification and recommendation of possible cooperation partners for IS exchange relationships

The visualization of the results is carried out by means of balance sheets, sankey diagrams, material cost matrices and sociograms. The functional requirements determined can be found in Table 1.

The first implementation step of the frontend component takes place with the input mask "Company", where the user enters the master data of the company. The location of the company can be selected from a dropdown menu, as the user can only access locations that have already been created. After all fields are filled in, the data can be transmitted to the backend via HTTP request by pressing the "save" button and then stored in the database.

The next step is the implementation of the "Materialmanagement" workspace. Here the user can create new material and energy inputs or outputs in the database as well as the material management of the company. Figure 2 shows the input mask for creating new processes with respective energy and material inputs/outputs in the database.

Using the buttons on the left side, the user can navigate between the work areas "Add Inputs", "Add Outputs" and "Management of processes and material". The company can be created as a black box[12] or in a detailed level with individual production lines and processes. Once the user decides to map the company with the internal structure, she/he is forwarded to the application area "Create production line". Here, individual production lines of the company can be created. The data of the production line consists of the name and a description. This data is created in the database after pressing the "save" button. Furthermore, a list of the created production lines is created on the page. By clicking on the corresponding button, processes can then be added to the production line. The production line can also be edited and deleted. If a production line is deleted that already contains process data, all linked data from the database are also deleted. Here the user can create a new process for the previously selected production line. The data of a process consists of the name, the production line, an item or index and the system costs incurred. By specifying an index, you can determine the position of the process in the system during evaluation using the MFCA method. As soon as a process has been created, inputs and outputs can be added to it. The inputs and outputs can be selected via a dropdown menu with a search function. The data for this are loaded from the database. At this point, the user can only access inputs or outputs that already exist in the database, so they need to be created beforehand. The selected inputs or outputs can then be supplemented with process-specific data. This includes, among other things, the quantity, the costs and the classification (i.e. the environmental impact, e.g. CO_2 emissions, of the input or output). In addition, it can be indicated whether the input or output is available

[12]Black box: the company is displayed as one process with all input and output flows.

Table 1 Overview of determined functional requirements

Nr.	User Administration (UA)
UA 1	To protect corporate data, only authenticated users should be able to use the application
UA 2	The administrator should be able to register users via a special form
UA 3	To use the application, the user must log in via the login window
Enterprise Administration (EA)	
EA 1	Every user should be able to create and edit specific data about the company
Material and Energy Management (MEM)	
MEM 1	Every user should be able to create new materials and energies (thermal and electrical) and store them in the database
MEM 2	Every user should have the possibility to create one or more production lines for the company
MEM 3	Every user should have the possibility to add one or more processes to existing production lines
MEM 4	Materials and energies can be added as input and output to created processes
MEM 5	Each material/energy should be able to be released by the user for exchange relationships
MEM 6	Individual product lines can be edited and deleted by the user
MEM 7	Individual processes can be edited and deleted by the user
Map functions (MF)	
MF 1	All companies in the industrial estate are to be displayed on a map
MF 2	It should be possible to filter the display of the companies on the map according to certain criteria: • Cooperation partners • Materials offered • geographical radius • Etc
Data Display and Visualization (DDV)	
DDV 1	Material and energy data should be presented in input–output balances
DDV 1.1	Input–output balances should be prepared for the entire company as well as for individual processes
DDV 1.2	All approved materials and energies (cross-company) should be presented in a site balance

(continued)

Table 1 (continued)

Nr.	User Administration (UA)
DDV 1.3	Particularly hazardous/non-hazardous materials should be color-coded
DDV 2	Material and energy data should be displayed in Sankey diagrams
DDV 2.1	Sankey diagrams should be created for the entire company as well as for individual processes
DDV 2.2	All approved materials and energies (cross-company) are to be displayed in a Sankey diagram for the site
DDV 2.3	All Sankey diagrams should represent both physical and monetary units of measure
DDV 3	Material and energy data from created production lines should be displayed in material cost matrices
DDV 4	Potential and actual cooperation links between companies should be presented in sociograms
Analysis Function (AF)	
AF 1	The data are to be evaluated according to the MFA method and thus material and energy movements in the company and location are to be determined
AF 2	The data are to be evaluated according to the MFCA method and thus cost saving potentials in the company or processes are to be pointed out
AF 3	Match-making algorithms will be used to determine possible cooperations between companies on the basis of material and energy data
Dashboard (DB)	
DB 1	Every user should see a dashboard with important information on the start page of the web application
DB 2	The dashboard should display data such as: • Material loss and use • CO_2 emissions • Possible symbiosis partners • Number of cooperation partners

for exchange relations with other enterprises. All entered energy and material data in physical units relate to the method of MFA and all cost-related data are required for MFCA.

A dashboard has been implemented on the start page of the application to provide an overview of the important key figures (Fig. 3). The dashboard shows, for example, possible cooperation partners, actual material losses and the company's CO_2 emissions. Starting from the start page, the user can navigate to all evaluations of the energy and material flows and the balance sheet. If the user has mapped the company with internal processes, process balances are displayed for each process created. A

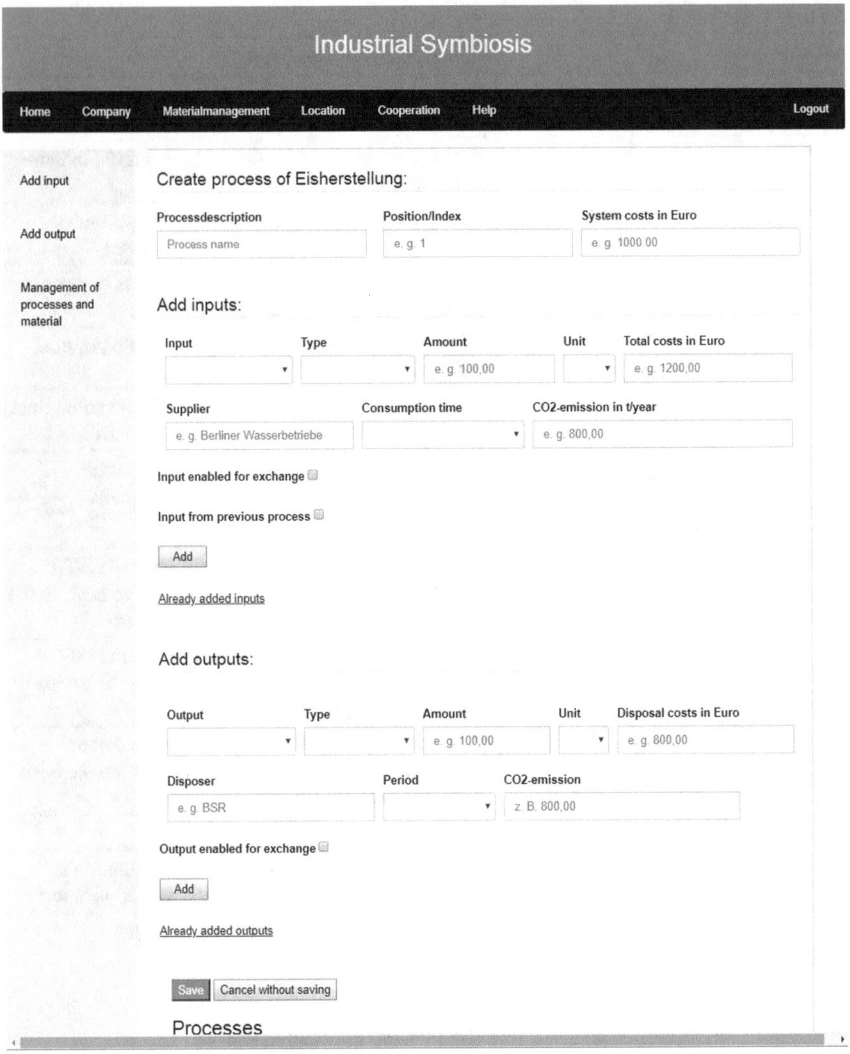

Fig. 2 Input mask for creating new processes with respective energy and material inputs/outputs

further representation of the material and energy evaluation was realized by implementing sankey diagrams,[13] representing quantity flows in physical and monetary units. By selecting the tab "Cost matrix" the user navigates to the MFCA evaluation (Fig. 4).

[13] A Sankey diagram is a graphical representation of quantity flows with arrows proportional to the quantity.

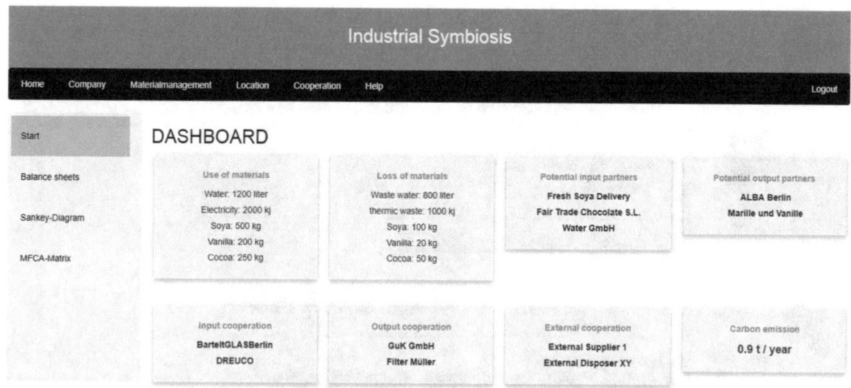

Fig. 3 Dashboard

Industrial Symbiosis

Home Company Materialmanagement Location Cooperation Help Logout

Start

Balance sheets

Sankey-Diagram

MFCA-Matrix

MFCA matrix

Vanilleeis

—	Material costs	Energy costs	System costs	Disposal costs	Σ
From previous process	0	0	—	—	0
New input	300	80	200	130	710
Total costs (process)	300	80	200	130	710
Product flow	217	56	140	—	413
Loss flow	83	24	60	130	[illegible]

Schokoladeneis

—	Material costs	Energy costs	System costs	Disposal costs	Σ
From previous process	50	0	—	—	50
New input	100	150	250	200	700
Total costs (process)	150	150	250	200	750
Product flow	127.5	105	175	—	407.5
Loss flow	22.5	45	75	200	[illegible]

Fig. 4 MFCA matrix

Another important implementation step is the workspace "cooperation". Possible cooperation partners are proposed on the first page of this web area. The potential symbiosis partners are determined in the backend component by corresponding queries of the database and transmitted to the frontend. For energy and material input–output resource matchmaking among the entities, the user can define a percentage value for the maximum deviation in quantity (Fig. 5). Potential exchange relationships that lie outside this limit are not displayed, so only companies that are within the critical quantity are shown. In the tab "Show own cooperations", a list of the own cooperation partners is displayed. This list includes the companies in the industrial area with which an exchange relationship already exists, which can additionally be

Industrial Symbiosis

Home Company Materialmanagement Location Cooperation Help Logout

Find cooperation

Show own cooperation

Show network diagram

Show area cooperation

Company : EisQueen GmbH

max. quantity deviations (%)

find

potential input cooperation partners

	Name	Input	Amount
1	EisQueen GmbH	Eiscreme	100 kg

potential output cooperation partners

	Name	Output	Amount
1	EisQueen GmbH	Eiscreme	120 kg

Fig. 5 Workspace "Cooperation"

represented by sociograms. On the one hand, a sociogram is created for connections to all companies, including external suppliers and disposal, and on the other hand for internal relationships within the industrial site.

Figure 6 shows the location function on the implemented leaflet map. In this work area the user can see all companies of the industrial area on the map by Default. Using the buttons on the left side, the user can select which markers are to be displayed, for example all cooperation partners or companies of the same branch. The last two buttons allow the display of companies based on required inputs or offered outputs. Markers are created by converting the stored addresses of the corresponding companies into geocoordinates (latitude and longitude) and transmitting them to the frontend.

Fig. 6 Map function

The web application allows on the one hand a company-specific analysis and on the other hand an analysis of the entire industrial site with methods such as MFA, MFCA and SNA. Every participating company can use the tool for internal evaluation and decision making and an established overarching organizational unit of the industrial park can use aggregated information for (resource/environmental) site management as well as the expansion of IS networks.

4 Discussion and Concluding Impulses for Future Research

This IS tool serves as an IS facilitating platform, providing transparency among market players and proposing potential cooperation partners according to selectable criteria (e.g. geographical radius, material properties, material quality, purchase quantity, delivery period). So this IS tool builds the technology-enabled environment for the processes of first screening of IS possibilities and initiation for further complex business-driven negotiations and agreements for long-term IS business relationships, regarding security of supply of resources, including possible (seasonal, temporal and qualitative) variability and fluctuations, medium to long-term agreements on price and quality.

Considering all entities as actors operating in a commercial market, the motivation of implementing IS activities is predominantly economy-driven, that is why this presented IS tool incorporates the method of MFCA to represent the economic values of all energy and material flows, including waste streams. This can lead to higher awareness of potential energy, material and cost savings, so that the (successive) process of reducing waste flows is given a higher priority. Furthermore, potential risks of for example misuse of information [40] or starting barriers were tackled in advance by taking respective precautions of specifying access and editing rights. For example, a company can exploit the web tool for internal resource evaluation and optimization reasons, but can adjust the tool settings so that they do not participate in the overarching level of IS detection.

In order to continuously monitor and control the progress of an IS development, an indicator system can be set up to evaluate the IS performance [28]. By defining goals for the IP, for example zero waste-, zero emission-, CO_2 neutral-park, the trajectory and corresponding (IS) measures can be compiled in a roadmap. So further research in the IS context should address the development of a (key performance) indicator system.

Additionally, the examination of recurring patterns of commonly exchanged materials, IS activities and IS structural formations opens up the use of Artificial Intelligence (AI) techniques [27]. As well as the complementary use of AI techniques can turn an IT-supported IS tool into a comprehensive and holistic instrument with which future scenarios and transformation paths from the actual system state to the desired future vision can be simulated [27, 29].

By the digitalization of data collection and processes, IPs may evolve into Smart Industrial Parks (SIP) with intelligence monitoring, information processing and risk prevention [23]. So Environmental Management Information Systems (EMIS) need to be established within IPs to provide a reliable data basis for comprehensive valuable information and substantiated decision making, so that companies can track their economic, social and ecological performance through advanced and intelligent ICT solutions.

Declaration of Interest The authors declare no conflict of interest.

References

1. Alkaya, E., Böğürcü, M., Ulutaş, F.: Industrial symbiosis in Iskenderun Bay: A journey from pilot applications to a national Program in Turkey. In: Conference paper, Proceedings of the conference SYMBIOSIS 2014 (2014)
2. Bain, A., Shenoy, M., Ashton, W., Chertow, M.: Industrial symbiosis and waste recovery in an Indian industrial area. Resour. Conserv. Recycl. **54**, 1278–1287 (2010)
3. Chertow, M.: Uncovering industrial symbiosis. J. Ind. Ecol. **11**(11), 11–30 (2007)
4. Chertow, M., Gordon, M., Hirsch, P., Ramaswam, A.: Industrial symbiosis potential and urban infrastructure capacity in Mysuru, India. Environ. Res. Lett. (2019) (in press). https://doi.org/10.1088/1748-9326/ab20e
5. Chertow, M.R.: Industrial symbiosis: literature and taxonomy. Ann. Rev. Energy Environ. **25**(1), 313–337 (2000)
6. Chertow, M.R.: Industrial symbiosis. Encyclopedia Energy **3**, 407–415 (2004)
7. Chertow, M.R.: Industrial ecology in a developing context. In: Clini, C., Musu, I., Gullino, M. (eds.) Sustainable Development and Environmental Management, pp. 1–19. Springer, Berlin (2008)
8. Crutzen, P.J.: Geology of mankind. Nature **415**, 23 (2002)
9. Cui, H., Liu, C., Côté, R., Liu, W.: Understanding the evolution of industrial symbiosis with a system dynamics model: a case study of Hai Hua industrial symbiosis, China. Sustainability **10**(3873) (2018). https://doi.org/10.3390/su10113873
10. Domenech, T., Bleischwitz, R., Doranova, A., Panayotopoulos, D., Roman, L.: Mapping industrial symbiosis development in Europe—typologies of networks, characteristics, performance and contribution to the Circular Economy. Resour. Conserv. Recycl. **141**, 76–98 (2019). https://doi.org/10.1016/j.resconrec.2018.09.016
11. Doménech, T., Davies, M.: The social aspects of industrial symbiosis: the application of social network analysis to industrial symbiosis networks. Prog. Ind. Ecol. **6**(1), 68–99 (2009)
12. Dong, L., Fujita, T., Zhang, H., et al.: Promoting low-carbon city through industrial symbiosis: a case in China by applying HPIMO model. Energy Policy **61**, 864–873 (2013). https://doi.org/10.1016/j.enpol.2013.06.084
13. Earley, K.: Industrial symbiosis: harnessing waste energy and materials for mutual benefit. Renewable Energy Focus **16**(4), 75–77 (2015). https://doi.org/10.1016/j.ref.2015.09.011
14. EEA, European Environment Agency: More from less—material resource efficiency in Europe. EEA report No 10/2016 Technical report European Environment Agency (2016)
15. Ehrenfeld, J., Gertler, N.: Industrial Ecology in Practice – The Evolution of Interdependence at Kalundborg. J. Ind. Ecol. **1**(1), 67–79 (1997)
16. Erkman, S., Van Hezik, C.: Eco-industrial parks in emerging and developing countries: Achievements, good practices and lessons learned, a comparative assessment of 33 cases in 12 emerging and developing countries. UNIDO Draft Report, United Nations Industrial Development Organizations, Vienna (2014)

17. Golev, A., Corder, G., Giurcob, D.P.: Industrial symbiosis in Gladstone: a decade of progress and future development. J. Clean. Prod. **84**, 421–429 (2014). https://doi.org/10.1016/j.jclepro.2013.06.054
18. Heeres, R.R., Vermeulen, W.J.V., et al.: Eco-industrial park initiatives in the USA and the Netherlands: first lessons. J. Clean. Prod. **12**, 985–995 (2004)
19. Herczeg, G., Akkerman, R., Hauschild, M.Z.: Supply Chain Management in Industrial Symbiosis Networks. PhD thesis, Technical University of Denmark, pp. 7–45 (2016)
20. Hewes, A.K., Lyons, D.I.: The humanistic side of eco-industrial parks: champions and the role of trust. Regional Stud. **42**, 1329–1342 (2008)
21. Isenmann, R.: Beitrag betrieblicher Umweltinformatik für die Industrial Ecology—Analyse von BUIS-Software-Werkzeugen zur Unterstützung von Industriesymbiosen. In: Gómez, J.M., Lang, C., Wohlgemuth, V. (eds.) IT-gestütztes Ressourcen- und Energiemanagement. Springer, Berlin, Heidelberg (2013). https://doi.org/10.1007/978-3-642-35030-6_37
22. Ismail, Y.: Industrial symbiosis at supply chain. Int. J. Bus. Econ. Law **4**(1), ISSN 2289-1552 (2014)
23. Li, B., Xiang, P., Hu, M., Zhang, C., Dong, L.: The vulnerability of industrial symbiosis: a case study of Qijiang Industrial Park, China. J. Clean. Prod. (2017). https://doi.org/10.1016/j.jclepro.2017.04.087
24. Lombardi, D.R., Laybourn, P.: Redefining industrial symbiosis. J. Ind. Ecol. **16**(1), 28–37 (2012)
25. Lütje, A., Möller, A., Wohlgemuth, V.: A Preliminary concept for an IT-Supported Industrial Symbiosis (IS) Tool Using Extended Material Flow Cost Accounting (MFCA) - Impulses for Environmental Management Information Systems (EMIS). In: Bungartz, H.-J., Kranzlmüller, D., Wohlgemuth, V. (eds.) Advances and New Trends in Environmental Informatics. Springer Nature, Switzerland AG (2018)
26. Lütje, A., Willenbacher, M., Möller, A., Wohlgemuth, V.: Enabling the Identification of Industrial Symbiosis (IS) through Information Communication Technology (ICT). In: Proceedings of the 52nd Hawaii International Conference on System Sciences (HICSS), pp. 709–719 (2019), ISBN: 978-0-9981331-2-6, https://hdl.handle.net/10125/59511
27. Lütje, A., Willenbacher, M., Engelmann, M., Kunisch, C., Wohlgemuth, V.: Exploring the system dynamics of Industrial Symbiosis (IS) with Machine Learning (ML) Techniques – A Framework for a Hybrid-Approach. In: Schaldach, R., Simon, K.-H., Weismüller, J., Wohlgemuth, V. (eds.) Advances and New Trends in Environmental Informatics - ICT for Sustainable Solutions. © Springer Nature Switzerland AG 2019, pp. 117–130 (2019). https://doi.org/10.1007/978-3-030-30862-9
28. Lütje, A., Wohlgemuth, V.: Tracking sustainability targets with quantitative indicator systems for performance measurement of industrial symbiosis in industrial parks. Admin. Sci. **10**(3), Special Issue: Industrial Ecology and Innovation (2020). https://doi.org/10.3390/admsci10010003
29. Lütje, A., Wohlgemuth, V.: Requirements engineering for an industrial symbiosis tool for industrial parks covering system analysis, transformation simulation and goal setting. Admin. Sci. **10**(1), Special Issue: Industrial Ecology and Innovation (2020). https://doi.org/10.3390/admsci10010010
30. Martin, M.: Industrial Symbiosis in the Biofuel Industry: Quantification of the Environmental Performance and Identification of Synergies. Dissertation No. 1507, Linköping Studies in Science and Technology (2013). ISSN: 0345-7524
31. Mirata, M.: Experiences from early stages of a national industrial symbiosis programme in the UK: determinants and coordination challenges. J. Clean. Prod. **12**, 967–983 (2004)
32. Notarnicola, B., Tassielli, G., Renzulli, P.A.: Industrial Symbiosis in the Taranto industrial district: current level, constraints and potential new synergies. J. Clean. Prod. **122**, 133–143 (2016). https://doi.org/10.1016/j.jclepro.2016.02.056
33. Pakarinen, S., Mattila, T., Melanen, M., Nissinen, A., Sokka, L.: Sustainability and industrial symbiosis—the evolution of a Finnish forest industry complex. Resour. Conserv. Recycl. **54**(12), 1393–1404 (2010). https://doi.org/10.1016/j.resconrec.2010.05.015

34. Sakr, D., El-Haggar, S., Huisingh, D.: Critical success and limiting factors for eco-industrial parks: global trends and Egyptian context. J. Clean. Prod. **19**, 1158–1169 (2011). https://doi.org/10.1016/j.jclepro.2011.01.001
35. Sokka, L., Lehtoranta, S., Nissinen, A., Melanen, M.: Analyzing the environmental benefits of industrial symbiosis. J. Ind. Ecol. **15**(1), 137–155 (2010)
36. Song, X., Geng, Y., Dong, H., Chen, W.: Social network analysis on industrial symbiosis: a case of Gujiao eco-industrial park. J. Clean. Prod. **193**, 414–423 (2018). https://doi.org/10.1016/j.jclepro.2018.05.058
37. Sterr, T., Ott, T.: The industrial region as a promising unit for eco-industrial development reflections, practical experience and establishment of innovative instruments to support industrial ecology. J. Clean. Prod. **12**, 947–965 (2004)
38. Ulhasanah, N., Goto, N.: Preliminary design of eco-city by using industrial symbiosis and waste co-processing based on MFA, LCA, and MFCA of cement industry in Indonesia. Int. J. Environ. Sci. Dev. **3**(6), 553–561 (2012). https://doi.org/10.7763/IJESD.2012.V3.285
39. Van Berkel, R., Fujita, T., Hashimoto, S., Fujii, M.: Quantitative assessment of urban and industrial symbiosis in Kawasaki, Japan. . Environ. Sci. Technol. **43**(5), 1271–1281 (2009). https://doi.org/10.1021/es803319r
40. Van Capelleveen, G., Amrit, C., Yazan, D.M.: A literature survey of information systems facilitating the identification of industrial symbiosis. In: Otjacques, B. et al. (eds.) From Science to Society, Progress in IS. In Springer International Publishing AG (2018). https://doi.org/10.1007/978-3-319-65687-8_14
41. Wallner, H.P.: Towards sustainable development of industry: networking, complexity and eco-clusters. J. Clean. Prod. **7**, 49–58 (1999)
42. Yu, F., Han, F., Cui, Z.: Evolution of industrial symbiosis in an eco-industrial park in China. J. Clean. Prod. **87**, 339–347 (2015). https://doi.org/10.1016/j.jclepro.2014.10.058
43. Yuan, Z., Shi, L.: Improving enterprise competitive advantage with industrial symbiosis: case study of a smeltery in China. J. Clean. Prod. **17**, 1295–1302 (2009). https://doi.org/10.1016/j.jclepro.2009.03.016
44. Zhu, Q., Lowe, E.A., Wei, Y., Barnes, D.: Industrial Symbiosis in China: a Case Study of the Guitang Group. J. Ind. Ecol. **11**, 31–42 (2008). https://doi.org/10.1162/jiec.2007.929

Deriving Benchmarks for Construction Products Based on Environmental Product Declarations

Anna Carstens, Tobias Brinkmann, and Barbara Rapp

Abstract The Life Cycle Assessment (LCA) results presented in Environmental Product Declarations (EPDs) are complex and difficult to interpret for many EPD owners. Benchmarks provide a way to aid the interpretation of EPDs by positioning the LCA results of a product in relation to the LCA results of comparable products. In this paper, benchmarking methods which have already been applied to results of LCAs are analysed according to their applicability to EPDs. The methods of data envelopment analysis (DEA) and determination of reference values are identified as being applicable to EPDs. For comparison, these methods are applied to a case study of insulation materials. A key difference in these methods lies in the fact that reference values are calculated for individual indicators while DEA calculates efficiency for every product as a single score.

Keywords Benchmark · Life cycle assessment · Environmental product declarations

1 Introduction

Environmental Product Declarations (EPDs) according to the norm EN 15804+A1 quantify the environmental impacts, resource use as well as output flows and wastes during a product's life cycle in 25 indicators. Basis for every EPD is a Life Cycle Assessment (LCA) according to the norms DIN EN ISO 14040/44. EPDs aim at communicating neutral and objective information rather than assessing whether a product is environmentally friendly or not. As a consequence, the LCA results of

A. Carstens (✉) · B. Rapp
University of Oldenburg, Oldenburg, Germany
e-mail: anna.carstens@brandsandvalues.com

B. Rapp
e-mail: b.rapp@uol.de

T. Brinkmann
brands & values GmbH, Bremen, Germany
e-mail: tobias.brinkmann@brandsandvalues.com

A. Kamilaris et al. (eds.), *Advances and New Trends in Environmental Informatics*, Progress in IS, https://doi.org/10.1007/978-3-030-61969-5_3

EPDs are complex and difficult to interpret for many EPD owners. This leads to a decreased interest in the instrument [1] and hinders EPD owners from using their EPDs to identify optimisation potentials from an environmental perspective. One issue in the communication of LCA results and thus EPDs is that, based on the LCA results in the EPDs, it is usually not possible to position a product among its competitors to find out whether it is a comparatively environmentally friendly product or not [2]. Therefore, a research gap regarding the "lack of positioning a product among its peers" is identified [2].

One approach to solve this problem is to determine benchmarks for LCA results [2]. Benchmarking is a management method whose aim is to identify and adapt improvement potentials through comparative procedures. One of the most common definitions of benchmarking is the identification of best practices which lead to a superior performance of an organisation or a product compared to its competitors [3]. Thus, benchmarking aims to not only determine how a company or product performs in comparison to other companies or products, but also to continuously improve a product's performance [4]. Procedures and solutions already applied by other companies should be identified and adapted in order to optimise own processes or products [5].

A benchmark thereby is a reference point or "best practice" against which a product's performance is measured, while benchmarking describes the process of determining this benchmark [5].

The fact that benchmarking of EPDs can be improved and that it is desired by EPD owners, is shown by a survey on the expectations and benefits of EPDs conducted by brands & values GmbH (b&v) in 2018 with more than 100 EPD stakeholders (EPD owners, EPD producers, programme operators and external auditors) from 15 countries. One finding of this survey is that, especially in the benefit category "benchmarking", the expectations of EPDs cannot be met [6]. The expectations for the benefit category "benchmark between products" were high on average, but the degree of fulfilment of this benefit category was rather low [6].

Benchmarking is used in various areas and so there is no universally applicable benchmarking method. In a literature review on the topic of "Benchmarks and LCA" a total of four benchmarking methods, which have previously been applied to LCA results, were identified [2]. As of spring 2020, it is not clear which of these benchmarking methods is suitable for benchmarking LCA results of EPDs, since it has not yet been investigated where the differences between the methods and their respective advantages and shortages lie [2]. The aim of this paper is to answer the questions "Which benchmarking methods can be used to derive benchmarks based on EPDs?" as well as "How and why do these methods differ in their results?". These questions will be investigated using the example of insulation materials as a case study for building products.

2 Methods

To answer the first question "Which benchmarking methods can be used to derive benchmarks based on EPDs?", the existing literature review by [2] is used as a starting point. Since this literature review only included articles up to the year 2017, a literature search for the years 2018 and 2019 was carried out. For this, the procedure of the original literature review was followed [2]. The databases SCOPUS and Web of Science were searched using the keywords "Life Cycle Assessment", "Life Cycle Analysis" or "LCA" as well as "Benchmark" or "Benchmarking", and all combinations of these keywords. However, since EPDs are created for products, only those papers which apply benchmarking methods to products (as opposed to organisations) are considered within the scope of this research. The identified papers were analysed to assess whether the applied methods can also be applied to the LCA results in EPDs.

To answer the second question "How and why do these methods differ in their results?", the methods which can be applied to EPD results are applied to a case study. For this case study, 19 insulation materials which can be applied for internal wall insulation were selected from the ÖKOBAUDAT database. The functional unit is defined as 1 m^2 of insulation material improving the U-value from 2.3 W/(m^2 K) to 0.15 W/(m^2 K). The modules A1-A3 over the impact categories global warming potential (GWP), ozone depletion potential (ODP), photochemical ozone creation potential (POCP), acidification potential (AP), eutrophication potential (EP), abiotic depletion for non-fossil resources (ADPE) and abiotic depletion for fossil resources (ADPF) were used for calculating the benchmarks.

To allow further comparison and easier communication of the benchmarks, the seven considered indicators are normalised, weighted and aggregated to a single score. In order to evaluate the extent to which the results agree, the linear correlation is calculated [7]. Since weighting factors are based on subjective values, no universally applicable or acceptable set of weighting factors can be defined. Therefore, three weighting methods were selected for this paper to show potential impacts on the resulting single scores and benchmarks. In the context of this paper, single scores of the following methods are compared: internal normalisation and equal weighting (IN), internal normalisation and PEF weighting [8] (IN+PEF) and b&v single score (b&v) [9]. For internal normalisation, the highest result of the respective environmental impact category is used as a normalisation factor. For the weighting, the PEF scaling factors [8] were scaled to reflect only the environmental impact categories under consideration. The b&v single score is determined according to [9].

3 Results

In total, 24 articles were identified for the literature review. Of those, 13 were identified in the previous literature review, five articles from year 2018 and six from 2019. The articles were according to [2] and are listed in Table 1.

3.1 Description and Selection of the Benchmarking Methods

In the following section, the identified methods are described, and it is evaluated whether these methods can be applied to EPDs. To be applicable to EPDs, a benchmarking method must be able to be applied solely on the basis of LCA results. A further requirement is that the method gives a definition of how to classify the "best practice".

Data envelopment analysis

Data Envelopment Analysis (DEA) determines the relative efficiency of so-called decision-making units (DMUs) using mathematical programming [10].

By multiplying inputs and outputs by so-called "significance weights", DEA makes it possible to compare inputs and outputs that are not initially comparable since they are measured in different units. This is relevant for LCA indicators, since the potential environmental impacts are communicated in different and non-convertible units. The results of the DEA are presented as a single score, which describes the

Table 1 Identified benchmarking methods (updated from [2])

Benchmarking method	Main objective	Adopted procedures	References
1. Data envelopment analysis (DEA)	Environmental efficiency assessment	3-step LCA+DEA	[7, 11–14]
2. Statistical analysis	Dimensionality reduction of variables	Principal component analysis, cluster analysis, linear and parametric correlations	[15–18]
	Determination of reference values	Mean, standard deviation, quartile analysis, mean+fuzzy numbers	[19–29]
3. Creation of indicators	Integrative assessment	Technical, social, environmental, financial and efficiency aspects	[30, 31]
	Single indicator assessment	Sensitivity analysis	[32]
4. Anchoring	Link results to a known reference	Relation with total daily impact	[33]

efficiency over all the input categories. Values of 1 are considered efficient in the context of a DEA and serve as a reference for the inefficient DMUs (efficiency <1).

DEA can be, and has already been, used to determine benchmarks based on EPDs [7]. Therefore, this method is also considered within the scope of this paper.

Following [7], an input-oriented model with variable returns to scales is used.

The environmental impacts can be interpreted as inputs for the production of a product.

Statistical analysis: Dimensionality reduction

One of the challenges for a benchmark is that LCAs, and thus also EPDs, contain a large number of indicators, some of which are redundant [16]. Dimensionality reduction aims to reduce the large number of indicators that are evaluated in a life cycle assessment or an EPD.

However, in the context of dimensionality reduction, no benchmarks in the sense of "best practice" are calculated, which is why this method is not considered within the scope of this paper.

Statistical analysis: Determination of reference values

A total of eight articles were identified, which establish reference values based on LCA results. The determination of reference values aims to quantify a benchmark using LCA data from comparable products. This can be done using different methods, which differ mainly in where a benchmark is set, i.e. which criteria of products have to be fulfilled in order to belong to "best practice". References [20, 22] calculate average values for residential buildings in Italy and Europe respectively. Whether these are benchmarks in the sense of "best practices" can be questioned, however, as the mean value represents the average.

A "best practice" can be defined, for example, on the basis of mean value and standard deviation [19, 29]. Collado-Ruiz and Ostad-Ahmad-Ghorabi [19] define products that perform better than most products on the market and can therefore be described as "best practice" as products whose environmental impact in the environmental impact category under consideration is less than the mean value minus two standard deviations. Under the assumption of a normal distribution, the best 2.4% of products considered meet this criterium.

An alternative approach for calculating reference values is quartile analysis. Here the first quartile, i.e. the best 25%, is defined as "best practice" and thus the target range, the third quartile as the limit range and the 50% in the middle as the reference range for the environmental impacts [21, 27, 28]. References [23–25] define the best 5% as an ambitious but feasible benchmark for residential buildings.

Reference values can be calculated for EPDs and are therefore considered in this paper. The quartile analysis is applied since this approach is used in particular in the construction industry in the valuation of buildings and is therefore relevant for EPDs [28].

Creation of indicators

This method is defined as incorporating aspects beyond LCA to create new indicators [2]. Therefore, these methods are not evaluated in this paper.

Anchoring

In anchoring, LCA results are compared to familiar products to provide a reference [33]. This method is similar to normalisation [2] and does not set benchmarks in the sense of best practices. Therefore, this method is not further considered in this paper.

3.2 *Application of Selected Benchmarking Methods*

In this section, the benchmarking methods DEA and reference values are applied to the 19 selected insulation materials from the ÖKOBAUDAT. It is important to note that these insulation materials were selected to demonstrate the applicability of the methods. No judgement about which insulation material is best is to be made from this study. Requirements for LCA studies making comparative statements and intended for publication are specified in [34]. For example, an assessment on the comparability of the studied systems carried out by interested parties as a critical review is required [34].

Data envelopment analysis (DEA)

For the implementation of DEA it is important that not too many variables are considered, otherwise too many DMUs can be classified as efficient [10]. Against this background, redundant indicators should first be removed. To determine redundant indicators, the linear correlation of the environmental categories is calculated. Indicators that do not correlate strongly with one another (correlation <0.9) are further considered [7]. The linear correlations are listed in Table 2.

GWP correlates strongly with the indicators AP and ADPF. Furthermore, there is a strong correlation between AP and EP. GWP is retained for the analysis because the indicator has high political significance and is widely known and applied. This

Table 2 Linear correlation between the environmental impact categories

	GWP	ODP	POCP	AP	EP	ADPE
ODP	−0.39					
POCP	−0.24	−0.14				
AP	**0.92**	−0.36	−0.21			
EP	0.90	−0.38	−0.32	**0.94**		
ADPE	0.33	−0.16	−0.04	0.31	0.36	
ADPF	**0.96**	−0.33	−0.08	0.86	0.81	0.36

implies that AP and ADPF are not considered further for the DEA. GWP and EP also correlate just under 0.9 (89.8%). However, both indicators are kept for the analysis.

DEA was carried out using the MultiplierDEA package [35] in RStudio [36].

The results of the DEA are presented in Table 3. Efficiency scores of 1 are efficient and thus form the benchmark. Products with an efficiency value <1 offer potential for improvement in relative comparison. In total, 6 of the 19 products are rated as efficient. The target values for the individual products can be calculated by using the lambdas of the efficient unit. The target value for GWP_{P3}, for example, is $0.9744 * GWP_{P15} + 0.0255 * GWP_{P19}$. Individual target values are calculated for each of the DMUs based on the efficient DMUs and the lambdas (see Table 3).

Statistical analysis: Reference values

Reference values for each of the seven considered environmental impact categories were calculated based on the quartile analysis. The best 25% in each impact category are defined. The results are presented in Table 4.

3.3 Comparison of Results

In this section, the results of the two methods are compared. A fundamental difference between the two methods is that reference values determined by statistical analysis are calculated for each environmental impact category whereas the DEA efficiency scores are presented in a single score.

Another difference between the methods is the number of products which are classified as the best practice and therefore the benchmark. For the reference values, all products in the first quartile can be seen as benchmarks. By definition, 25% of the products are therefore classified as belonging to the best practice. With DEA however, the number of efficient and therefore benchmark products will vary depending on the inputs.

When reference values are calculated, the target inputs equal to the first quartile for every impact category are applicable to all the considered products. Therefore, one set of target values is calculated. For the DEA, however, the target values differ for each DMU as shown in Table 3.

Table 5 shows the ratio of the DEA target values to the reference values based on the 25% quartile. No general statement can be made about which target values are more ambitious. For the impact category GWP, the DEA target values for 17 of the 19 products are lower and therefore more ambitious than the reference values. In comparison, the DEA target values for EP are lower and therefore more ambitious than the reference values for only seven of the 19 products. This is because for ten of the DEA inefficient products, the product P15 determines the target value more than 95% (see Table 3). In the impact category GWP, the product P15 belongs to the best 25%. In the impact category EP, however, the product P15 does not belong to the best 25%. Hence the DEA target values are higher and therefore less ambitious than the reference value in for the impact category EP.

Table 3 Results of the DEA

DMU	Efficiency	Lambdas						Targets				
		P1	P2	P7	P11	P15	P19	GWP (kg CO_2-eq)	ODP (kg CFC-11-eq)	POCP (kg Ethen-eq)	EP (kg PO_4^{3-}eq)	ADPE (kg Sb-eq)
P1	1	1	0	0	0	0	0	1.30E+01	7.44E−08	9.75E−02	2.73E−03	5.15E−06
P2	1	0	1	0	0	0	0	1.31E+01	6.81E−08	1.07E−01	2.72E−03	5.04E−06
P3	0.32	0	0	0	0	0.9744	0.0255	1.29E+01	8.97E−11	4.28E−03	6.71E−03	2.79E−06
P4	0.48	0	0	0	0	0.9749	0.025	1.28E+01	8.82E−11	4.25E−03	6.72E−03	2.79E−06
P5	0.62	0	0	0	0	0.9758	0.0241	1.28E+01	8.56E−11	4.21E−03	6.72E−03	2.78E−06
P6	0.77	0	0	0	0	0.9744	0.0256	1.29E+01	9.00E−11	4.28E−03	6.71E−03	2.79E−06
P7	1	0	0	1	0	0	0	3.05E+01	3.43E−06	3.45E−03	2.39E−02	2.57E−06
P8	0.39	0	0	0	0	0.9673	0.0326	1.29E+01	1.10E−10	4.64E−03	6.70E−03	2.84E−06
P9	0.69	0.1068	0	0	0.8693	0.0238	0	1.43E+01	1.85E−05	1.89E−02	4.72E−03	2.06E−05
P10	0.94	0	0	0	0.9351	0.0557	0.0092	1.44E+01	1.99E−05	9.68E−03	5.01E−03	2.17E−05
P11	1	0	0	0	1	0	0	1.45E+01	2.13E−05	9.65E−03	4.91E−03	2.30E−05
P12	0.86	0.0131	0	0	0.9771	0.0098	0	1.44E+01	2.08E−05	1.07E−02	4.90E−03	2.25E−05
P13	0.25	0	0	0	0	1	0	1.26E+01	1.62E−11	2.99E−03	6.77E−03	2.61E−06
P14	0.41	0	0	0	0	1	0	1.26E+01	1.62E−11	2.99E−03	6.77E−03	2.61E−06
P15	1	0	0	0	0	1	0	1.26E+01	1.62E−11	2.99E−03	6.77E−03	2.61E−06
P16	0.15	0	0	0	0.0045	0.9831	0.0124	1.28E+01	9.58E−08	3.65E−03	6.74E−03	2.79E−06
P17	0.26	0	0	0	0.0045	0.9831	0.0124	1.28E+01	9.58E−08	3.65E−03	6.74E−03	2.79E−06
P18	0.22	0	0	0	0.0001	0.9573	0.0427	1.30E+01	2.27E−09	5.15E−03	6.68E−03	2.91E−06
P19	1	0	0	0	0	0	1	2.09E+01	2.90E−09	5.35E−02	4.58E−03	9.78E−06

Table 4 Reference values

	GWP (kg CO_2-eq)	ODP (kg CFC-11-eq)	POCP (kg Ethen-eq)	AP (kg SO_2-eq)	EP (kg PO_4^{3-}-eq)	ADPE (kg Sb-eq)	ADPF (MJ)
Minimum	1.26E+01	1.62E−11	2.99E−03	2.97E−02	2.72E−03	2.57E−06	1.11E+02
1. Quartile	1.67E+01	8.08E−10	8.11E−03	4.59E−02	5.49E−03	8.11E−06	3.58E+02
3. Quartile	6.31E+01	2.04E−06	2.41E−02	1.59E−01	2.24E−02	1.69E−04	8.44E+02
Maximum	1.34E+02	3.28E−05	1.07E−01	5.12E−01	4.58E−02	3.68E−03	1.99E+03

Table 5 Ratio of DEA target values and reference values (values < 1 indicating that the reference value is higher than the DEA target are highlighted yellow)

Product	GWP	ODP	POCP	EP	ADPE
P1	0.78	92.01	12.02	0.50	0.63
P2	0.78	84.18	13.22	0.49	0.62
P3	0.77	0.11	0.53	1.22	0.34
P4	0.77	0.11	0.52	1.22	0.34
P5	0.77	0.11	0.52	1.22	0.34
P6	0.77	0.11	0.53	1.22	0.34
P7	1.83	4239.28	0.43	4.36	0.32
P8	0.77	0.14	0.57	1.22	0.35
P9	0.85	22900.22	2.33	0.86	2.54
P10	0.86	24623.07	1.19	0.91	2.68
P11	0.87	26331.98	1.19	0.89	2.83
P12	0.86	25730.18	1.32	0.89	2.78
P13	0.76	0.02	0.37	1.23	0.32
P14	0.76	0.02	0.37	1.23	0.32
P15	0.76	0.02	0.37	1.23	0.32
P16	0.76	118.56	0.45	1.23	0.34
P17	0.76	118.56	0.45	1.23	0.34
P18	0.78	2.81	0.64	1.22	0.36
P19	1.25	3.58	6.60	0.83	1.21

For further comparison with DEA, the EPD results were normalised, weighted and aggregated to a single score. This procedure has already been suggested and applied by [7] for comparing DEA results. Single scores of the following methods are compared: internal normalisation and equal weighting (IN), internal normalisation and PEF weighting [8] (IN+PEF), b&v single score (b&v) [9] and DEA. The resulting weighting factors are listed in Table 6.

Table 7 shows the single score results of the methods considered and the ranking of the results. For better visualisation, the lowest 25% of the values are coloured green, the highest 25% red and the middle 50% yellow, analogous to the reference values.

Table 6 Applied weighting factors

Indicator	Equal weighting (%)	b&v (%)	PEF (%)
GWP	14.29	41.25	32.9
ODP	14.29	5.58	10.0
POCP	14.29	5.58	7.5
AP	14.29	15.18	9.9
EP	14.29	15.18	14.7
ADPE	14.29	2.05	11.9
ADPF	14.29	15.18	13.2

Table 7 Comparison of single scores

	IN	Rank IN	IN + PEF	Rank IN+PEF	b&v	Rank b&v	1-DEA	Rank DEA
P15	0.06	1	0.07	1	0.09	1	0.00	1
P19	0.17	5	0.15	5	0.17	7	0.00	1
P11	0.17	6	0.15	4	0.14	2	0.00	1
P7	0.18	8	0.21	11	0.24	12	0.00	1
P1	0.19	9	0.14	2	0.14	3	0.00	1
P2	0.20	11	0.15	3	0.14	4	0.00	1
P10	0.18	7	0.16	7	0.15	5	0.06	7
P12	0.20	10	0.17	8	0.16	6	0.14	8
P6	0.13	2	0.16	6	0.19	8	0.23	9
P9	0.27	15	0.23	12	0.20	9	0.31	10
P5	0.16	4	0.20	10	0.23	11	0.38	11
P4	0.21	12	0.26	13	0.31	13	0.52	12
P14	0.16	3	0.18	9	0.22	10	0.59	13
P8	0.26	14	0.34	15	0.41	15	0.61	14
P3	0.32	16	0.40	16	0.46	16	0.68	15
P17	0.34	17	0.41	17	0.49	17	0.74	16
P13	0.25	13	0.29	14	0.36	14	0.75	17
P18	0.51	18	0.55	18	0.54	18	0.78	18
P16	0.61	19	0.73	19	0.88	19	0.85	19

On the basis of the singles scores and the rankings, similarities between the methods can be identified. P15 is ranked first in all methods. P18 and P16 occupy 18th or 19th rank. However, there are also differences in the methods. For IN results, the product P6 is in second place. If the other methods are applied, P6 only ranks sixth (IN+PEF) to ninth (DEA). This is due to the fact that P6 has relatively low potential environmental impacts in the impact categories ODP and POCP. However, ODP and POCP have relatively low weights in both the b&v and the PEF weighting, so that the relatively low potential environmental impacts have less influence on the single score result than in the IN. It can therefore be concluded that the weighting factors impact the single score and therefore the ranking of the different products, especially when the products have relatively low emissions in some impact categories and relatively high emissions in other impact categories.

The linear correlation of the results is between 0.99 (IN+PEF and b&v) and 0.72 (IN and DEA) (see Table 8). In particular, the correlation between DEA and the other methods is relatively low.

Table 8 Linear correlation of the results

	IN	IN+PEF	DEA
IN+PEF	0.97		
DEA	0.72	0.82	
b&v	0.93	0.99	0.83

Table 9 Weights DEA

Product	GWP	ODP	POCP	EP	ADPE
P1	6.55	0.00	0.24	2.54	0.00
P2	2.99	0.00	0.00	11.94	0.00
P3	0.00	0.11	0.22	2.14	0.00
P4	0.00	0.17	0.33	3.20	0.00
P5	0.00	0.22	0.42	4.13	0.00
P6	0.00	0.27	0.52	5.14	0.00
P7	0.12	0.00	0.00	0.00	1396.42
P8	0.00	0.14	0.26	2.58	0.00
P9	4.50	0.00	0.16	1.74	0.00
P10	0.00	0.33	0.64	6.26	0.00
P11	0.08	0.35	0.66	6.59	0.00
P12	5.66	0.00	0.20	2.19	0.00
P13	0.00	930.24	0.00	1.70	0.00
P14	0.00	1498.09	0.00	2.75	0.00
P15	2.18	113.76	0.24	5.33	0.00
P16	0.00	0.05	0.10	0.98	0.00
P17	0.00	0.09	0.18	1.76	0.00
P18	0.00	0.08	0.15	1.47	0.00
P19	0.00	0.33	0.66	6.63	3.19

This raises the question of which weighting factors are applied in DEA. The corresponding weights from the DEA are shown in Table 9. These were calculated with internally normalised indicator results to allow relative comparison.

It is noticeable here that a large number of the weights are equal to 0, so that these inputs are not taken into account for the calculation of efficiency [37]. This should be viewed critically, since the inputs were selected beforehand and should therefore also be taken into account in the efficiency calculation [38]. Furthermore, it is noticeable that the weighting of the same criteria of different DMUs varies greatly.

This issue can be overcome at least partly by integrating weight restrictions into the DEA model [10]. This has already been applied to LCA results [11]. To illustrate the effects of weight restriction, the ranking of weights by PEF is applied to DEA ($GWP \geq EP \geq ADPE \geq ODP \geq POCP$). In order to prevent the weights from diverging extremely and environmental indicators being weighted with zero, the constraint that an indicator may be weighted a maximum of three times as high as the following indicator. The results are then compared to the single score obtained by applying IN+PEF. As Fig. 1 shows, the results in terms of the ranking have been aligned. Although there are still some products (P6 and P9) which have different ranks according to the methods. This is due to the remaining flexibility in the weights with DEA.

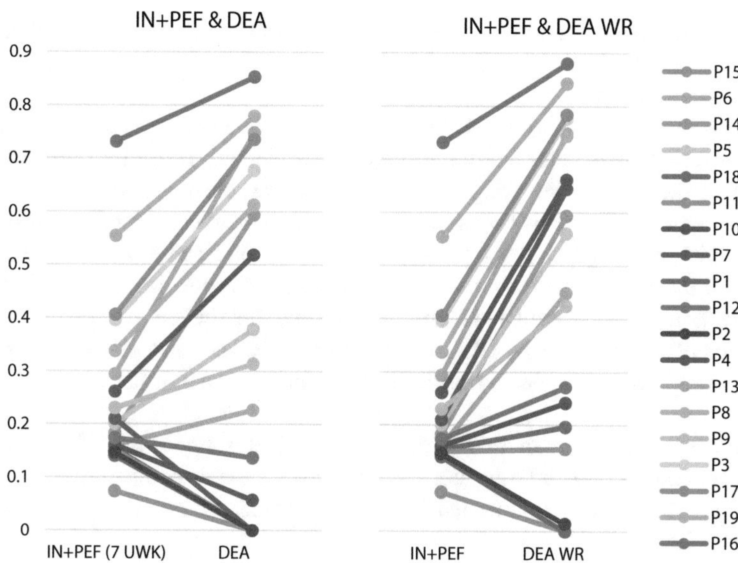

Fig. 1 Comparison of IN+PEF, DEA and DEA with weight restrictions (DEA WR)

4 Discussion and Outlook

This paper has demonstrated that the benchmarking methods determination of reference values by statistical analysis and DEA can be applied to derive benchmarks based on EPDs and highlighted key differences between the methods. Eventually, benchmarks aim at improving the communication of the EPD results to non-experts such as EPD owners. This poses the question of which of these methods is better suited at communicating EPD results to non-experts.

One key difference is that reference values are determined for each indicator while the DEA forms a single score. In a study with consumers in Finland, most potential users preferred a benchmark that weights the results of the different environmental impacts and aggregates them into a single score, as this presentation format is easy to understand [33]. Since not all persons are prepared to leave the weighting of environmental impacts to anonymous experts, the communication of unweighted and weighted benchmarks is recommended [33]. This could be achieved by communicating reference values for every environmental impact category as well as a single score.

A disadvantage of traditional weighting methods is, however, that subjective and/or normative evaluations are required to determine the weights. DEA has the advantage that decision-makers do not have to make subjective decisions in advance about the weighting of the individual environmental impacts considered, since the weights are determined within the DEA model. Nevertheless, the analysis has shown that this can lead to individual environmental impact categories not being considered at all or that the weighting of individual categories varies greatly between the considered products.

When comparing the results of the DEA to other single score methods, the analysis has shown that the differences in the single score methods result from the different weighting factors. For the weighting of LCA results, clear communication of the applied weighting factors is important in order to be able to consider the assumptions under which the single scores were calculated in the interpretation of the results. It is argued that the formation of a single score according to traditional weighting methods is more transparent and therefore more suitable for communication than the formation of a single score by DEA. Further, the EPD owner may define their own set of weighting factors to reflect priorities in, for example, the industry or geographical area.

The methods shown in this paper are aimed at supporting EPD owners to create benchmarks for internal use. LCA studies making comparative assertions and intended for publication need to fulfil a number of requirements to ensure comparability [34].

References

1. Kägi, T., Dinkel, F., Frischknecht, R., Humbert, S., Lindberg, J., De Mester, S., Ponsioen, T., Sala, S., Schenker, U.W.: Session "Midpoint, endpoint or single score for decision-making?"—SETAC Europe 25th Annual Meeting, May 5th, 2015. Int. J. Life Cycle Assess. **21**, 129–132 (2016). https://doi.org/10.1007/s11367-015-0998-0
2. Galindro, B.M., Zanghelini, G.M., Soares, S.R.: Use of benchmarking techniques to improve communication in life cycle assessment: a general review. J. Clean. Prod. **213**, 143–157 (2019). https://doi.org/10.1016/j.jclepro.2018.12.147
3. Camp, R.C.: Benchmarking—The Search for Industry Best Practices that Lead to Superior Performance. ASQC Quality Press, Wisconsin (1989)
4. Elmuti, D., Kathawala, Y.: An overview of benchmarking process: a tool for continuous improvement and competitive advantage. Benchmarking Qual.Manag. Technol. 4, 229-243 (1997). https://doi.org/10.1108/14635779710195087
5. Siebert, G., Maßalski, O., Kempf, S.: Benchmarking: Leitfaden für die Praxis. Carl Hanser Verlag GmbH & Co. KG, München (2008)
6. Brinkmann, T., Köhler, S., Boeth, A., Metzger, L.: Environmental Product Declarations - Benefits, Expectations and Fulfilments - A Stakeholder View - Part 1 of 2. https://www.brandsandvalues.com/study1-epd-environmentalproductdecl (2019)
7. Galindro, B.M., Bey, N., Olsen, S.I., Fries, C.E. and Soares, S.R.: Use of data envelopment analysis to benchmark environmental product declarations—a suggested framework. Int. J. Life Cycle Assess (2019). https://doi.org/10.1007/s11367-019-01639-1
8. European Commission: PEFCR Guidance document, - Guidance for the development of Product Environmental Footprint Category Rules (PEFCRs), version 6.3, December 2017 (2018)
9. Brinkmann, T., Metzger, L.: Ecological assessment based on environmental product declarations. In: Teuteberg, F., Hempel, M., Schebek, L. (eds.) Progress in Life Cycle Assessment 2018, pp. 21–31. Springer International Publishing, Cham (2019)
10. Cooper, W.W., Seiford, L.M., Tone, K.: Data Envelopment Analysis: A Comprehensive Text With Models, Applications, References and DEA-Solver Software. Springer, New York (2007)

11. Tatari, O., Kucukvar, M.: Eco-efficiency of construction materials: data envelopment analysis. J. Constr. Eng. Manage. **138**, 733–741 (2012). https://doi.org/10.1061/(ASCE)CO.1943-7862.0000484
12. Li, Q.J., Wang, K.C.P., Cross Stephen, A.: Evaluation of Warm Mix Asphalt (WMA): a case study. Airfield and Highway Pavement 2013, pp. 118–127. https://doi.org/10.1061/9780784413005.011
13. Iribarren, D., Marvuglia, A., Hild, P., Guiton, M., Popovici, E., Benetto, E.: Life cycle assessment and data envelopment analysis approach for the selection of building components according to their environmental impact efficiency: a case study for external walls. J. Clean. Prod. **87**, 707–716 (2015). https://doi.org/10.1016/j.jclepro.2014.10.073
14. Gonzalez-Garay, A., Guillen-Gosalbez, G.: SUSCAPE: A framework for the optimal design of sustainable chemical processes incorporating data envelopment analysis. Chem. Eng. Res. Des. **137**, 246–264 (2018). https://doi.org/10.1016/j.cherd.2018.07.009
15. Curzons, A.D., Jiménez-González, C., Duncan, A.L., Constable, D.J.C., Cunningham, V.L.: Fast life cycle assessment of synthetic chemistry (FLASC^{TM}) tool. Int J Life Cycle Assess. **12**, 272 (2007). https://doi.org/10.1065/lca2007.03.315
16. Gutiérrez, E., Lozano, S., Adenso-Díaz, B.: Dimensionality reduction and visualization of the environmental impacts of domestic appliances. J. Ind. Ecol. **14**, 878–889 (2010). https://doi.org/10.1111/j.1530-9290.2010.00291.x
17. Ji, C., Hong, T., Jeong, J., Kim, J., Lee, M., Jeong, K.: Establishing environmental benchmarks to determine the environmental performance of elementary school buildings using LCA. Energy Build. **127**, 818–829 (2016). https://doi.org/10.1016/j.enbuild.2016.06.042
18. Genovese, A., Morris, J., Piccolo, C., Koh, S.C.L.: Assessing redundancies in environmental performance measures for supply chains. J. Clean. Prod. **167**, 1290–1302 (2017). https://doi.org/10.1016/j.jclepro.2017.05.186
19. Collado-Ruiz, D., Ostad-Ahmad-Ghorabi, H.: Comparing LCA results out of competing products: developing reference ranges from a product family approach. J. Clean. Prod. **18**, 355–364 (2010). https://doi.org/10.1016/j.jclepro.2009.11.003
20. Moschetti, R., Mazzarella, L., Nord, N.: An overall methodology to define reference values for building sustainability parameters. Energy Build. **88**, 413–427 (2015). https://doi.org/10.1016/j.enbuild.2014.11.071
21. Gervasio, H., Dimova, S., Pinto, A.: Benchmarking the life-cycle environmental performance of buildings. Sustainability **10**, 1454 (2018). https://doi.org/10.3390/su10051454
22. Lavagna, M., Baldassarri, C., Campioli, A., Giorgi, S., Dalla Valle, A., Castellani, V., Sala, S.: Benchmarks for environmental impact of housing in Europe: Definition of archetypes and LCA of the residential building stock. Build. Environ. **145**, 260–275 (2018). https://doi.org/10.1016/j.buildenv.2018.09.008
23. Hollberg, A., Lützkendorf, T., Habert, G.: Using a budget approach for decision-support in the design process. In: Presented at the IOP Conference Series: Earth and Environmental Science (2019)
24. Hollberg, A., Lützkendorf, T., Habert, G.: Top-down or bottom-up? How environmental benchmarks can support the design process. Build. Environ. **153**, 148–157 (2019). https://doi.org/10.1016/j.buildenv.2019.02.026
25. Hollberg, A., Vogel, P., Habert, G.: LCA benchmarks for decision-makers adapted to the early design stages of new buildings **8** (2018)
26. Chen, Y., Ng, S.T., Hossain, M.U.: Approach to establish carbon emission benchmarking for construction materials. Carbon Manage. **9**, 587–604 (2018). https://doi.org/10.1080/17583004.2018.1522094
27. Rasmussen, F.N., Ganassali, S., Zimmermann, R.K., Lavagna, M., Campioli, A., Birgisdóttir, H.: LCA benchmarks for residential buildings in Northern Italy and Denmark – learnings from comparing two different contexts. Build. Res. Inf. **47**, 833–849 (2019). https://doi.org/10.1080/09613218.2019.1613883
28. Paratscha, R., Von Der Thannen, M., Smutny, R., Lampalzer, T., Strauss, A., Rauch, H.P.: Development of LCA benchmarks for Austrian torrent control structures. Int. J. Life Cycle Assess. **24**, 2035–2053 (2019). https://doi.org/10.1007/s11367-019-01618-6

29. Pasanen, P., Castro, R.: Carbon Heroes Benchmark Program—Whole building embodied carbon profiling. IOP Conf. Ser.: Earth Environ. Sci. **323**, 012028 (2019). https://doi.org/10.1088/1755-1315/323/1/012028
30. Zea Escamilla, E., Habert, G.: Global or local construction materials for post-disaster reconstruction? Sustainability assessment of twenty post-disaster shelter designs. Build. Environ. **92**, 692–702 (2015). https://doi.org/10.1016/j.buildenv.2015.05.036
31. Rönnlund, I., Reuter, M., Horn, S., Aho, J., Aho, M., Päällysaho, M., Ylimäki, L., Pursula, T.: Eco-efficiency indicator framework implemented in the metallurgical industry: part 1—a comprehensive view and benchmark. Int. J. Life Cycle Assess. **21**, 1473–1500 (2016). https://doi.org/10.1007/s11367-016-1122-9
32. Gül, S., Spielmann, M., Lehmann, A., Eggers, D., Bach, V., Finkbeiner, M.: Benchmarking and environmental performance classes in life cycle assessment—development of a procedure for non-leather shoes in the context of the product environmental footprint. Int. J. Life Cycle Assess. **20**, 1640–1648 (2015). https://doi.org/10.1007/s11367-015-0975-7
33. Nissinen, A., Grönroos, J., Heiskanen, E., Honkanen, A., Katajajuuri, J.M., Kurppa, S., Mäkinen, T., Mäenpää, I., Seppälä, J., Timonen, P., Usva, K., Virtanen, Y., Voutilainen, P.: Developing benchmarks for consumer-oriented life cycle assessment-based environmental information on products, services and consumption patterns. J. Clean. Prod. **15**, 538–549 (2007). https://doi.org/10.1016/j.jclepro.2006.05.016
34. ISO 14044: DIN EN ISO 14044:2018-05, Umweltmanagement – Ökobilanz – Anforderungen und Anleitungen (ISO 14044:2006 + Amd 1:2017); Deutsche Fassung EN ISO 14044:2006 + A1:2018 (2018)
35. Aurobindh Kalathil Puthanpura: MultiplierDEA: Multiplier Data Envelopment Analysis and Cross Efficiency. R package version 0.1.18., https://CRAN.R-project.org/package=MultiplierDEA
36. RStudio Team: RStudio: Integrated Development for R. RStudio, Inc., Boston, MA. https://www.rstudio.com/
37. Ströhl, F., Borsch, E., Souren, R.: Integration von Gewichtsrestriktionen in das DEA-Modell nach Charnes, Cooper und Rhodes : Exemplarische Optionen und Auswirkungen. Ilmenauer Schriften zur Betriebswirtschaftslehre. 3/2018, 36 (2018)
38. Roll, Y., Cook, W.D., Golany, B.: Controlling factor weights in data envelopment analysis. IIE Trans. **23**, 2–9 (1991). https://doi.org/10.1080/07408179108963835

Sustainability

Circular Economy in the Rostock Region. A GIS and Survey Based Approach Analyzing Material Flows

Ferdinand Vettermann, Samer Nastah, Laurine Larsen, and Ralf Bill

Abstract Within this work a method how to analyze the recycling sites in Rostock and the surrounding municipalities is presented. The approach is embedded into the project Prosper-Ro and is part of the development of a decision support system for planning purposes in the mentioned region. The recycling sites are one part of the project and in focus to tackle the challenge of transforming the linear economy to a circular economy. To make the method easy adoptable, it incorporates free available data, like Open Street Map and governmental data, such as ALKIS, to derive a high resolution dataset for the population density as well as the potential waste. This leads to an easy adoptable approach for other municipalities. With Open Street Map and QGIS routing software QNEAT it was possible to create reachability areas for each recycling site in respect to the customer and waste potential. These data lead to a parameter of 250 customers per 1 m^2 of container area in the city area of Rostock which is useful for planning new sites or extending the existing ones. The derived data could be verified by the help of a survey, which shows that the customers in Rostock are mostly satisfied with their recycling sites and want to improve the transformation to an circular economy, for instance by the reuse of electrical devices.

Keywords Circular economy · Rostock · Survey · GIS

F. Vettermann (✉) · S. Nastah · R. Bill
University of Rostock, Justus-von-Liebig-Weg 6, 18057 Rostock, Germany
e-mail: ferdinand.vettermann@uni-rostock.de

S. Nastah
e-mail: samer.nastah@uni-rostock.de

R. Bill
e-mail: ralf.bill@uni-rostock.de

L. Larsen
BN Umwelt GmbH, Petridamm 26, 18046 Rostock, Germany
e-mail: l.larsen@bn-umwelt.de

A. Kamilaris et al. (eds.), *Advances and New Trends in Environmental Informatics*, Progress in IS, https://doi.org/10.1007/978-3-030-61969-5_4

1 Introduction

1.1 Project Background

This work is embedded into the project Prosper-Ro (Prospective synergistic planning of development options in the regiopole using the example of the urban–rural area of Rostock) [1]. The project is funded by the BMBF (Federal Ministry of Education and Research) and involves several partners in the region of Rostock. The goal is to develop a decision support system based on ecosystem services for Rostock and the surrounding communities. The system is still under development, but the presented work will be an important element. One part of the project as well as for the decision support is the analysis of the material flows between the city part and the rural areas. For this purpose, the main partner is the BN-Umwelt GmbH. The material flows are focused onto the circular economy, especially the waste management with recycling facilities. These facilities are embedded into the German waste bringing system.

To make proposals how the waste management in relation to the bringing system can be improved, a survey as well as a GIS analysis were conducted. One goal is to analyse how much waste could be expected and how the inhabitants in the regiopole can as fast as possible get to the recycling areas. This survey also assesses the quality of the recycling sites. This is important to give the customers a better experience to improve their will to bring their waste to the disposal sites. Furthermore the survey is aimed to detect the potential to take or bring functioning electric devices. This could reduce the amount of electronic waste as well as the need for new products. This problem is also addressed by the European Union. For instance, the customers should get a right to repair of electronic devices [2]. This could help to reduce CO^2-emissions. The EU-commission assumed, that with a proper circular economy, the CO^2-emissions in the EU can be reduced by 48% until 2030 and up to 83% until 2050 [3]. Also, the conversion to a circular economy could lead to 700 000 new jobs as well as a rise in the GDP by 0.5% until 2030 [2]. In order to help achieving this goal, this project aims for improvements on a local scale.

1.2 Situation of the Circular Economy in Germany

As part of the German waste management system, the users of the facilities need to bring their waste to guarantee a proper treatment. This is enforced by the circular economy law (Kreislaufwirtschaftsgesetz) [4], with the four main goals:

1. waste prevention,
2. recycling,
3. waste disposal and
4. other waste management methods.

There are differences between the rural areas around Rostock (the administrative district Rostock, abbreviated as LRO) and the city of Rostock (the district-free city Rostock, named HRO) expectable. The most important difference is the reachability of the recycling sites with respect to the area which has to be supplied. Furthermore, the type of waste would be different as a result of a varying settlement structure [5]. In HRO, the sum of separately collected waste for recovery per inhabitant is 192 kg, for LRO it is below 178 kg per inhabitant in 2018 [6]. The residual waste in HRO is around 290 kg/inhabitant and 158 kg/inhabitant in LRO [6]. HRO is therefore above the average (482 kg/inhabitant vs. 462 kg/inhabitant), whereas LRO with 336 kg/inhabitant is significantly under the average of the Federal Republic of Germany [7].

The main difference between urban and rural areas can be found in the waste fractions. In the cities, the part of green cut is mostly lower than in rural areas [5]. Furthermore, glass bottles are mostly substituted through plastic bottles in cities [8]. Also, the fraction of packaging waste in the populated areas is higher than in the rural areas. And the most important fact is that the engagement of the inhabitants to separate their waste is much lower in the cities than in rural areas. A big problem, however, is the comparability of the data. The municipalities have different fee statutes for the different waste fractions [5]. For instance, the delivery of green cut in HRO is free while in LRO a fee has to be paid. This difference also leads to a different treatment by the customers. For example, some choose to compost their green cut on their own or to dump it illegally [9].

Another cause of the different waste fraction treatments is related to the age of the residents [10]. The age is also highly correlated with the actual housing situation and the household size. The higher amount of waste in cities mostly results from the bigger part of single person households [11].

To involve the different settlement structure, the most important part is the differentiation between one- or two-family houses and households in multi-family houses. Surprisingly, there is no big difference between the amount of house and bulky waste. Furthermore, better educated inhabitants are producing a lower amount of waste. In contrast the amount of waste of unemployed inhabitants seems to be higher [9].

2 Methodology

2.1 Gathering Spatial and Waste Related Data

One of the most important parts of the project is the incorporation of as much openly available data as possible. For HRO opendata.hro as a portal for all kinds of spatial data is available. It includes, for example, actual facts of inhabitants and the social structure for each part of the city. For LRO the situation for free spatial data is less comfortable, but there are also some sources available. To get information about the population situation in each municipality, the data portal of the GDI-DE (spatial infrastructure for whole Germany) becomes handy. To get an overview of existing open data, portal.opengeoedu.de is a good point to start with [12].

The calculation of reachability is performed on the basis of a traffic network based on Open Street Map (OSM). Therefore, the different highway types are extracted for the area of interest around Rostock. To calculate the reachability of the different sites in a high precision, each road needs a specific maximum speed attribute. This attribute is calculated by the median of the available maximum speeds for each highway type. In comparison to the mean value, the median offers "normal" maximum speed values (e.g. 50 km/h). Furthermore it seems that the network with median values is slower than with mean values. For the calculation of the reachability this offers a margin of safety, because it seems better for the customer to reach the site faster than proposed.

For the calculation of the user potential of each site, the population density as well as the waste potential is calculated on the base of the actual land use. The land use dataset is based on the German ALKIS system (Official Real Estate Cadastre Information System). This system incorporates different kind of land use. This information gives the opportunity to estimate a potential population density as well as a waste density for each land use type (Table 1). The ALKIS dataset is the only dataset which is not freely available in Mecklenburg-Western Pomerania in contrast to other federal states in Germany (e.g. Hamburg, Berlin, Rhineland-Palatine, Thuringia). Nevertheless it is an important data source to adapt the workflow for other municipalities in Germany. The population densities in Table 1 are assigned manually through a discussion process with BN-Umwelt, which has much expertise in this kind of research. The waste densities are based on the survey which is shown later on as well as on empirical values [9, 13]. The waste density involves the higher

Table 1 Land use classes with population and waste density

Code	Description	Population density in %	Waste density in %
2000	Closed residential area	100	80
1000	Open residential area	80	100
2110	Living with public	80	50
2120	Living with trade and services	80	50
2130	Living with business and industry	80	50
2730	Living with business	80	50
2140	Public with living	50	25
2150	Trade and services with living	50	25
2160	Business and industry with living	50	25
2100	Buildings and open space, mixed use with residential	50	25
4310	Weekend and holiday home area	10	50
4440	Allotment garden	10	80
1460	Accomodation	10	10

percentage of green cut, which mostly contributes to the waste fractions at the recycling sites. On the other hand, the high population density in the closed residential area also contributes to a high amount of waste. Furthermore, the gardening areas are also weighted higher. The reason for this is also the fraction of green cut.

To obtain the information about the recycling sites itself, for HRO construction plans for digitalization were available. For LRO the sites are digitized using aerial images. This leads to a lower accuracy and provides a less detailed information about the sites itself. For example, in HRO the information about the size and the type of garbage for each container is available, in LRO only a rough number for the container area can be calculated (Fig. 1).

Information about the waste itself in HRO is available from the year 2018 and later on. This dataset is provided by the recycling sites itself. These data is collected

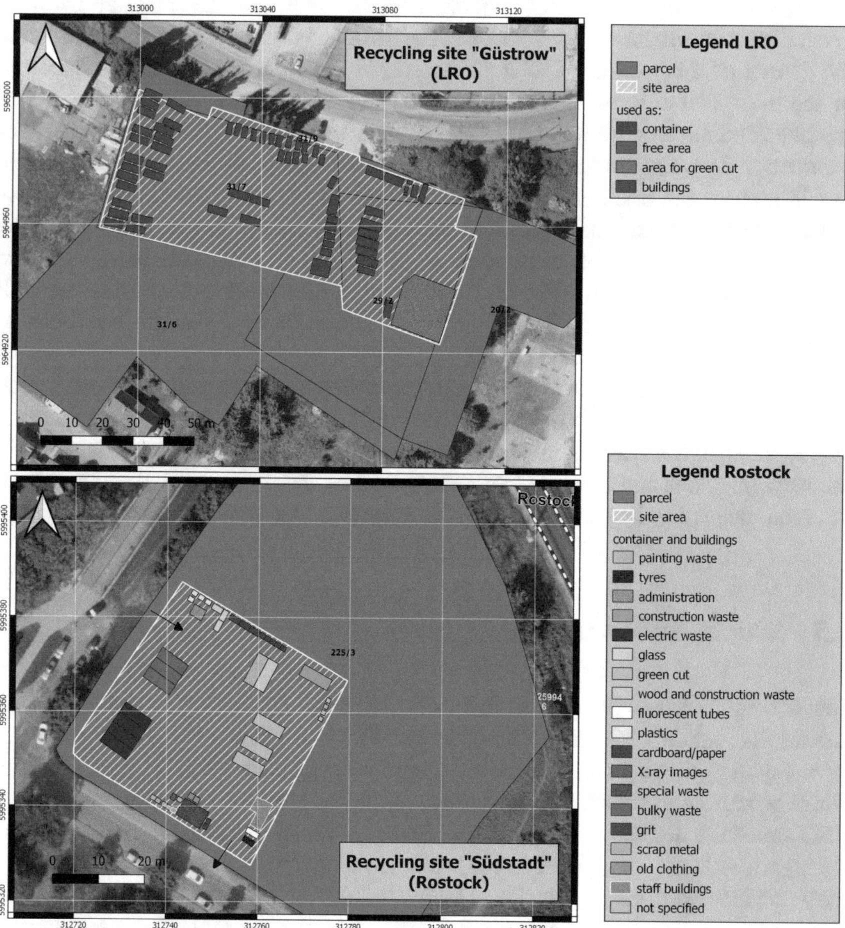

Fig. 1 Example of two recycling sites and the differences between HRO and LRO

from each customer who is delivering waste. This information has been analyzed for the year 2018 and is very useful to compare it with the analysis within this work [14].

2.2 Survey

One part within this work was the development of a survey, which is aimed at gathering information about the users of the recycling sites. The survey was designed as guided interview at the sites itself and as an online survey. The survey "Recycling site of the future" was created with EvaSys, a web tool for creating online surveys that also offers the possibility to get a statistical conclusion of the collected information.

The survey contains four different categories of questions. The first part includes general user information, like postcode. The postcode is used to locate and visualize the source of the waste. Of interest is also the age and the residential situation of the user. This information is important to evaluate a connection to the waste fractions and the amount of waste. A valuable information is also if the person has a garden to derive possible parameters. The second part of the survey is focused on the different waste fractions which are brought by the user and about the readiness to bring and take functional, electrical devices. The third part is about the ways how users inform themselves about the recycling sites and how they reach the site (by car, with public transportation). The fourth part is about the customer satisfaction with the services and the reachability of the recycling site. There is also the possibility to give improvement recommendations.

All this information is used to derive possible parameters and to detect, which factors are connected and how much waste could be expected. In the future, another survey campaign in LRO is planned to show differences and common features in the recycling sites and how they are recognized by the users. Also, the focus is on customer needs and how the site could be expanded to a service center.

2.3 Spatial Analyses

One of the most important parts of this work are the calculations of reachability areas and the amount of potential waste and users. With this information, the potential for new recycling sites as well as an expansion of existing sites could be proposed. The whole workflow is shown in Fig. 2. First of all, the areas of the same reachability (ISO areas) are derived with the QGIS plugin QNEAT3. Further information about the algorithm itself is available in [15]. Due to the fact that the users have to pay for the waste they bring to sites outside of their administrative districts, the borders of the city Rostock are treated as hard borders. This leads to a low amount of cross border waste [14]. To achieve this, the calculation for HRO and LRO is made separately. This should lead to a worse reachability in the border areas. The ISO areas are classified

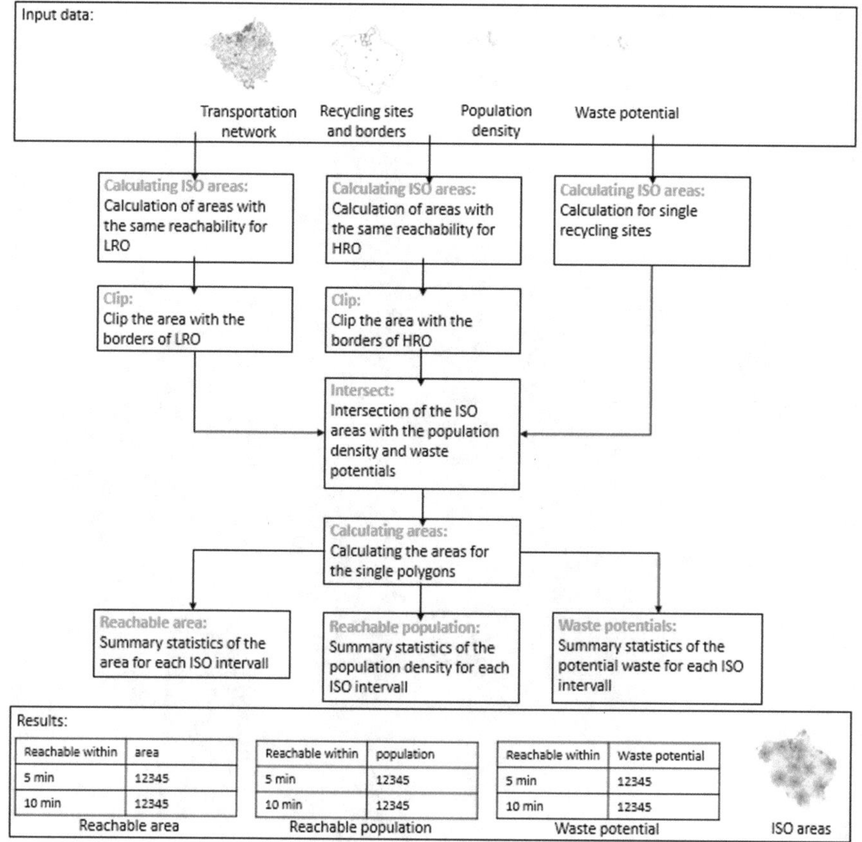

Fig. 2 GIS workflow for the derivation of the reachability regions

in 5 min steps (<5 min; 5–10 min; 10–15 min; 15–20 min; 20–25 min; 25–30 min; >30 min). Furthermore, one ways were handled as normal roads in relation to the lower data quality of OSM.

The next step is the intersection of the ISO areas with the potential waste and population density. Furthermore, this calculation is done for each single recycling site in HRO. This leads to an assumption of the study area coverage as well as potential amounts of waste and users for the single recycling sites.

3 Results

The reachability of the recycling sites in HRO and LRO is evaluated as good (Fig. 3). 98% of the population could reach the facilities within 15 min in HRO and 71% in LRO. Under the assumption of a longer acceptable way of 20 min to the nearest

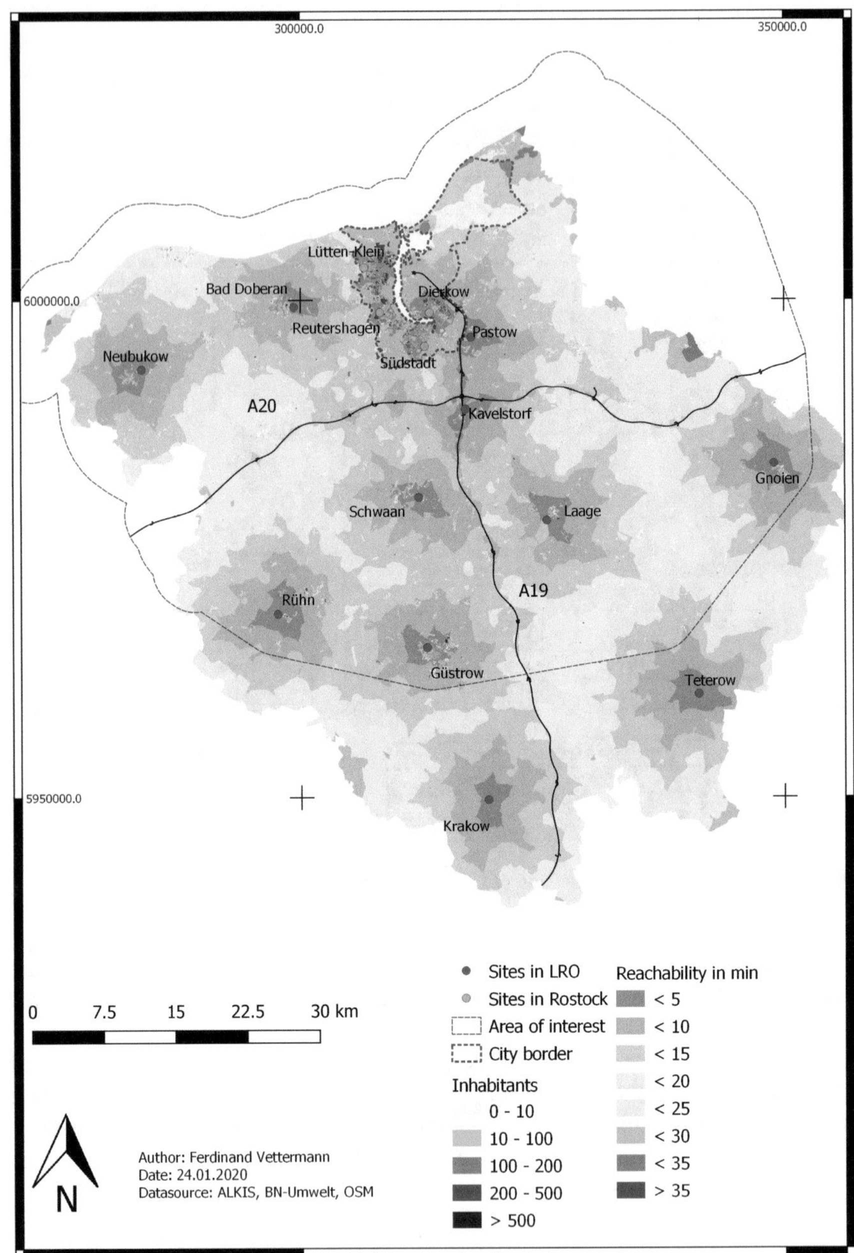

Fig. 3 Reachability of the recycling sites in HRO and LRO

Table 2 Waste potential, customer potential and possible parameters for the recycling sites in Rostock

Site	Customer potential	Waste potential	Site area in m^2	Factor site area	Traffic area in m^2	Factor traffic area	Container area in m^2	Factor container area
Lütten Klein	85,534	72,512	5752	13	4539,08	16	530	137
Dierkow	90,315	83,917	4378	19	3844,4	22	369	227
Südstadt	72,485	80,551	1941	42	1607,78	50	282	286
Reutershagen	82,135	88,748	2293	39	1871,24	47	373	238

recycling site, the part of the reachable inhabitants in LRO rises to 87%. In HRO, 68% of the city area and 60% of the area of LRO are reachable within the 15 min ISO area.

The distribution over the whole area shows that there are several parts with a lower reachability. This concerns the area in the north east, in the western part around the motorway A20 and the low populated areas in the south. The amount of inhabitants that can reach a recycling center in 25 min and more is in LRO 4% and in HRO < 1%. Furthermore, at the border between HRO and LRO, a drop in the supply is visualized. This is also the case for the border regions of LRO to other counties.

To evaluate the status of the single recycling sites, the number of covered people as well as the waste potential was calculated. Due to the lack of data in LRO, this is only processed for the city of Rostock (Table 2). To obtain the information about the capacity, the area of the recycling sites was used for getting a rough amount of possible users. The customer potential and waste potential are derived by the amount of reachability zone 1 (5 min), 50% of zone 2 (10 min) and 10% of zone 3 (15 min). This concerns the users who are able to move between different sites. From these results possible parameters are derived. The parameters describe the amount of waste which could be stored but also more important is the factor of the permeability of the users—how fast they are able to check in, unload their waste and leave the site. Therefore, the area used for traffic seems mostly reasonable. But on closer investigation this factor varies strongly from one site to another. Under the assumption that only two of the sites are temporary congested which are getting enhanced in the future, the factor is without significance.

Also the complete area of the site offers nothing more valuable. The only factor which provides comparable results is the container area. Only the recycling site in "Lütten Klein" offers much more container space for each potential user in comparison to the other sites. The "Südstadt" site is the one that will be enhanced next, so the high customer amount per container area will most likely drop after the expansion. Under these general conditions, a parameter of 250 customers per m^2 container area seems reasonable. This means that the recycling sites in Rostock, except "Südstadt", still have potential for more users (Table 2).

To support these outcomes, the results of the survey comes into play. Over two months 202 evaluable questionnaires were completed. Unfortunately, the survey is not as representative as needed. Especially the age distribution is problematic. 37.1% of the questioned customers are 45–65 years old and 38.6% are >65 years old. In contrast, the group of the 15–25 years old customers contributes only 2%. Furthermore, the spatial distribution shows that the majority of customers are using the nearest recycling site. This supports the calculation of potential users. The problem of the hard borders between the city and rural areas is also identifiable in the survey: only 2% of the customers are from outside of HRO.

Overall, the customers in HRO are satisfied with the quality of the sites. 68% of the interviewed customers have no wishes for improvement. Also the reachability of the sites seems the most important element of the sites. Because of the good reachability, the situation in HRO seems very well. Between the sites there is no difference identifiable. This satisfaction with the sites supports also the chosen factor of 250 potential customers per m^2 container area.

One of the most important parts of the survey is the type of waste in combination with the residential situation of the customer. The main waste fraction was green cut (80%), electric waste (74%), bulky waste (58%) and painting waste (36%). In Fig. 4 it is visible, that green cut is mainly delivered by single family households. This also contributes for electric waste. The only waste fraction where the single family houses do not contribute the majority is the bulky waste. Therefore apartment buildings are the main source. The assumption beforehand, that the big residential buildings, in cause of the lower social status of their residents, are causing more waste, could not be verified just in regard to the recycling sites. It seems that they are contributing mostly domestic waste or they are not using the bringing system this intensively. In the derivation of parameters for the waste contribution, the big apartment buildings are still a major factor because of the high population density.

Moreover, 71% of the interviewed customers are contributing functional electric devices as waste. 68% of the interviewed customers are willing to take functioning electronic devices. To address these wishes, improvements have to be made to enhance the sites to service centres.

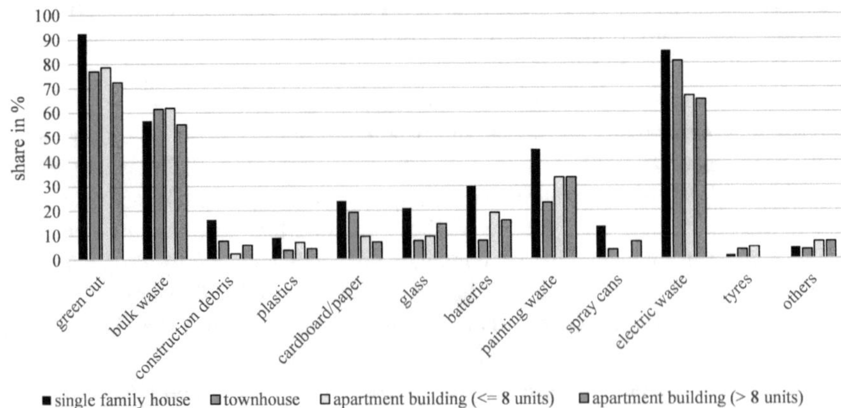

Fig. 4 Waste distribution of the different housing distributions

4 Discussion

In the following part, the method as well as the presented results will be discussed critically. Especially the calculation of the ISO areas could be improved with other methods and data. The traffic infrastructure is the bottleneck of the method, particularly the quality of OSM as well as the derived speed information. The crowdsourcing approach of OSM leads, for instance, to errors in the topology, the situation and the highway type [16]. These errors could be ruled out or at least reduced with commercial road infrastructure datasets, like HERE. On the other hand this would contradict the usage of mostly freely available data sources within this work. Moreover, the incompleteness of the speed information could be solved with the method shown above. Furthermore, the use of public transport could be involved. But the survey show that 90% of the customers are using the car to bring waste, only 2% of them are using public transports. This leads to the assumption, that the connection to public transport stations seems not necessary.

The population density and the potential waste amount are based on ALKIS. This leads to three potential errors. The first one is that the data in ALKIS does not necessarily have to correspond to the actual status due to the binding to changes in the land register. Furthermore, the density factors are derived manually based on experience in cooperation with BN-Umwelt. These parameters are not analysed in detail due to the lack of data. This will be changed in the following years due to the digitalization of the recycling sites in HRO as well as in LRO [17]. With this, a statistical approach will become possible. The third problem affects the accuracy of the area of the density polygons from ALKIS. They are not based on single buildings, so green areas around buildings are also estimated as populated. For single family houses as well as townhouses this is not problematic, however for bigger buildings it is. To improve the area accuracy building shapes and their floor count could be incorporated [18].

Another point to discuss it the permeability of the recycling sites. The chosen parameter model seems to be relatively simple in relation to the potential influence factors. For instance, rush hours are not taken into consideration. For instance, this applies to Saturday afternoons, when the majority of customers are delivering their waste. Furthermore, there is a seasonality in the amount of green cut. Especially during spring and autumn the amount will be higher [14]. Actually, it is only possible to create factors with a margin of safety which are involving these rush hours.

From the survey, it is difficult to derive a statement about the differences between the waste fractions of the single sites. The costumer only has to answer which fraction he is delivering and not how much waste it is. In contrast to the data of 2018, “Dierkow” is the site with the highest amount of green cut. Between the other sites, no big differences are visible. The lowest amount of green cut in the year 2018 was delivered to the recycling site “Südstadt” [14]. The area of allotment gardens in the vicinity of this site is also lower in comparison to the other sites. This leads to the assumption that the survey was not able to provide reasonable information about the waste amounts. This is only possible with the digital recycling site data.

Unfortunately, statements about social elements like age could not be derived. It was only possible to get information about the housing situation in combination with the waste fractions.

5 Conclusion

The method shown within this paper makes it possible to visualize and evaluate the waste flows in Rostock and the rural areas around the city. The whole work provides a workflow, how decisions about planning and extensions of recycling sites could be supported. This could easily be integrated into the decision support system being developed in the project Prosper-Ro. Furthermore, the work presented shows potential design parameters for the recycling sites in Rostock. In combination with the survey valuable information about the waste distribution as well as the contributors of waste was derived. Also, the survey was able to discover the overall satisfaction of the inhabitants of Rostock with their recycling sites. With this, potential factors to achieve the target to establish a circular economy in Europe are identified, for instance the readiness of the customers to take functional electric devices from the recycling sites.

With the derivation of reachability areas and their intersection with the population density as well as the potential waste, customer potentials can be determined. Especially the high resolution of these datasets should be mentioned. They are very useful input for other analyses besides the circular economy. Furthermore the open data provides an easy adaptable workflow for regions beside Rostock and the surrounding rural areas.

In the future, different enhancements are conceivable. This concerns especially for the survey which will be conducted in LRO to get comparable information for LRO as well. Furthermore, the parameter for the sites could be further specified, for instance for single waste fractions. With this information, statements about the capacity for the single waste fractions could be derived. The last step of the workflow will be the automation within a python script, to offer this service embedded into the web app which will be constructed through Prosper-Ro.

References

1. Tränckner, J.: PROSPER-RO: Prospektive synergistische Planung von Entwicklungsoptionen in Regiopolen am Beispiel des Stadt-Umland-Raums Rostock (2017)
2. Becker, M.: Brüssel will Reparaturpflicht für Elektrogeräte (2020). https://www.spiegel.de/wirtschaft/soziales/eu-will-reparaturpflicht-fuer-elektrogeraete-a-00000000-0002-0001-0000-000169828696?sara_ecid=soci_upd_wbMbjhOSvViISjc8RPU89NcCvtlFcJ. Accessed 09 Mar 2020
3. Ellen MacArthur Foundation, McKinsey Center for Business and Environment: Growth Within: A circular economy vision for a competetive Europe, Isle of Wight (2015)

4. Bundesministerium für Umwelt, Naturschutz und nukleare Sicherheit (BMU): Gesetz zur Förderung der Kreislaufwirtschaft und Sicherung der umweltverträglichen Bewirtschaftung von Abfällen: KrWG (2012)
5. Schlitte, F., Schulze, S.: Siedlungsabfallaufkommen in Deutschland. Wirtschaftsdienst **94**(9), 680–682 (2014). https://doi.org/10.1007/s10273-014-1733-3
6. Landesamt für Umwelt, Naturschutz und Geologie Mecklenburg-Vorpommern (2019) Daten zur Abfallwirtschaft 2018
7. Destatis (2020) Aufkommen an Haushaltsabfällen: Deutschland, Jahre, Abfallarten. https://www-genesis.destatis.de/genesis/online?sequenz=tabelleErgebnis&selectionname=32121-0001&zeitscheiben=2. Accessed 06 Mar 2020
8. Welle, F.: PET in Beverage Packagin. Kunststoffe Int. **98**(10), 106–109 (2008)
9. Buchert, M., Bleher, D., Dehoust, G. et al.: Demografischer Wandel und Auswirkungen auf die Abfallwirtschaft: Ermittlung der Auswirkungen des demografischen Wandels auf Abfallanfall, Logistik und Behandlung und Erarbeitung von ressourcenschonenden Handlungsansätzen (2017)
10. Hoffmeister, J.: Demografie und Abfall: Wechselwirkungen zwischen sozio-demografischen Einflussfaktoren und dem spezifischen Abfallaufkommen. In: Versteyl, A., Thomé-Kozmiensky, K.J. (eds.) Planung und Umweltrecht, vol. 2. TK Verl. Karl Thome-Kozmiensky, Neuruppin (2008)
11. Erichsen, J.-O., Schlitte, F., Schulze, S.: Entwicklung und Determinanten des Siedlungsabfallaufkommens in Deutschland. Hamburg Institute of International Economics (HWWI) (2014)
12. Hinz, M., Bill, R.: Ein zentraler Einstiegspunkt für die Suche nach offenen Geodaten im deutschsprachigen Raum. AGIT – Journal für Angewandte Geoinformatik (4): 298–307 (2018)
13. Bilitewski, B., Härdtle, G., Marek, K.: Abfallwirtschaft: Handbuch für Praxis und Lehre. Springer: Berlin, Heidelberg (2000)
14. Stegert, M.: Nutzerorientierte Bedarfsplanung zur Ermittlung des Optimierungspotenzials von Recyclinghöfen. Bachelorarbeit (2019)
15. Raffler, C.: QNEAT3: IsoArea Algorithms (2018). https://root676.github.io/IsoAreaAlgs.html. Accessed 01 Jul 202
16. Kirchmayer-Novak, S.: Bewertung der Datenqualität der OpenStreetMap als Datengrundlage für einen Verkehrsgraph mit Hilfe offener Daten und Software: am Anwendungsbeispiel Einzugsbereiche von Park-and-Ride Anlagen im südlichen Wiener Umland. Masterarbeit (2014)
17. Ostseezeitung: Wertstoffhöfe im Landkreis werden digital. https://www.ostsee-zeitung.de/Mecklenburg/Bad-Doberan/Wertstoffhoefe-im-Landkreis-werden-digital (2019). Accessed 24 Jan 2020
18. LaIV: Gebäudemodelle. https://www.laiv-mv.de/Geoinformation/Geobasisdaten/gebaeude%E2%80%93modelle/ (2019). Accessed 24 Jan 2020

Obsolescence as a Future Key Challenge for Data Centers

Fabian A. Schulze, Hans-Knud Arndt, and Hannes Feuersenger

Abstract The advance of sustainable development in times of climate change, energy revolution and emerging scarcity of resources has become a new global challenge, which gives organizations the responsibility to make processes more sustainable. Obsolescence as a product aging process with complex causalities is unavoidable and, in the context of sustainability considerations, the question arises how to deal effectively with obsolescence along the entire product life cycle. Especially in the area of data center operation, a significant growth in capacity has been recorded in recent years, with organizations using cloud computing solutions to reduce space and IT equipment and to outsource them to centralized data centers, which increases the challenges for data centers in dealing with obsolescence in a sustainable manner. In cooperation with the data center Biere in Saxony-Anhalt, an as-is state regarding the management of obsolescence has been recorded. It was found that obsolescence management approaches are already being implemented in operational data centers. However, there is still potential for optimization regarding sustainable development when dealing with obsolescence, especially in procurement processes, in product life cycle management, and in cooperation with hardware vendors and service providers.

Keywords Sustainability · Obsolescence management · Data centers

1 Introduction

In recent years, there has been an increase in computing power and energy requirements in data centers. A data center is understood to be an "organizational unit that offers computing and support activities in a centralized manner and has powerful

F. A. Schulze · H.-K. Arndt (✉) · H. Feuersenger
Otto-Von-Guericke-Universität, Universitätsplatz 2, 39106 Magdeburg, Germany
e-mail: hans-knud.arndt@iti.cs.uni-magdeburg.de

F. A. Schulze
e-mail: fabian.schulze@st.ovgu.de

H. Feuersenger
e-mail: hannes.feuersenger@ovgu.de

A. Kamilaris et al. (eds.), *Advances and New Trends in Environmental Informatics*, Progress in IS, https://doi.org/10.1007/978-3-030-61969-5_5

hardware and software systems at its disposal" [1]. Previous sustainability-oriented considerations of data centers have mainly taken into account the increased energy consumption.[1] This has led to the development of performance measurement systems monitoring energy efficiency in data centers [3]. However, based on these observations, data on so-called grey energies in upstream and downstream activities of the systems and the technology used in the data center are rarely recorded. For example, the extraction and processing of natural resources such as gold, silver, copper, neodymium and tantalum used in IT infrastructure can be mentioned [4]. Nevertheless, in the context of sustainable development, it is crucial to take a holistic view on how the technology used in the data center is handled from the aspects of durability and sustainability.

Green IT[2] has led to an increase in the demand for sustainable IT infrastructure [6] and customers want IT solutions—as part of their own value creation—to be more sustainable. In discussions about research areas of Green IT, data centers are often implicitly mentioned when taking a closer look at all interacting components [7].

Products (and also services) are subject to obsolescence. This term is used to describe "the aging of a product in terms of its production method, materials or other factors, which makes it obsolete or unusable" [8]. There are many reasons for obsolescence, and dealing with obsolescence as capacity grows, both quantitatively and qualitatively, presents new challenges for data center management. Exemplary are the reduction of negative environmental impacts and the efficient use of limited natural resources in the technology used in the data center as well as the cost-efficient use of data center infrastructure.

Since obsolescence is unavoidable [10], data centers are not faced with the challenge of whether it is handled effectively, but rather how. With regard to sustainable development, this paper discusses the management of obsolescence on the basis of an as-is analysis of a large data center. Furthermore, it will be shown how data centers can advance sustainable development in dealing with obsolescence.

2 Increasing Importance of Data Centers

In accordance with DIN EN ISO 14001:2015 on environmental management, a data center is an independent organization with "its own functions and its own administration" [9] consisting of data center infrastructure—i.e. the physical building with security facilities, power supply, emergency systems, air conditioning up to racks and cables—and the IT infrastructure, the IT equipment required for the provision of services, mostly servers, storage, routers, switches, and so on, and therefore a

[1]For the increase in energy demand from data centers of over 25% in the period from 2010 to 2017, see [2].

[2]Green IT "refers to the resource-conserving use of energy and input materials in information and communication technology over the entire life cycle, i.e. that already at the development stage, not only the most resource-conserving use of the technology in operation is taken into account, but also environmentally friendly disposal and reuse of the input materials." [5].

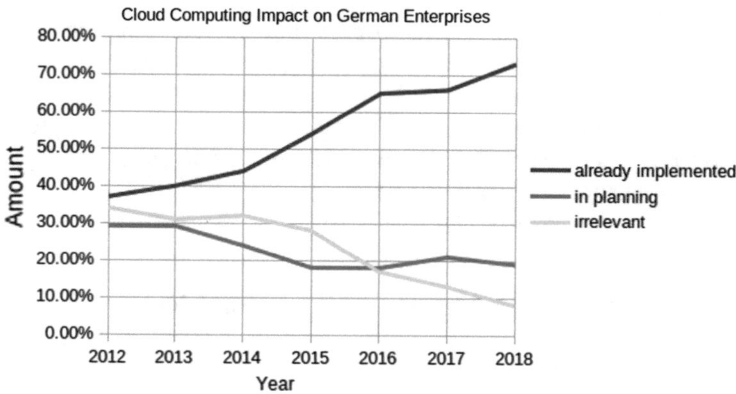

Fig. 1 Relevance of cloud computing for German companies (own representation based on the data from [13, 14, 17, 18])

not inconsiderable amount of software and hardware products [11]. Distinctions between data centers can be made with regard to the characteristics of performance, location or the degree of automation. New innovative data center projects are often created by combining these characteristics in new ways. In general, a distinction can be made between company-owned computer centers (so-called in-house computer centers) and external computer centers whose operators primarily provide services for customers (so-called service data centers). Whereas the owner of in-house data centers are responsible for the trouble-free operation of the data center themselves and thus also for IT equipment and premises, in external data centers the responsibility for IT equipment and premises lies with the provider of the data centers.

The data center of our project partner T-Systems located in Biere in Saxony-Anhalt is one of the largest data centers in Europe [12]. The range of services offered is characterized by a percentage of traditional outsourcing solutions of almost 50%, whereas the number of cloud computing services is growing constantly. Figure 1 provides an overview of the relevance of cloud computing technology for German companies. Whereas in 2012 only 37% of German companies used cloud computing technology for their processes, this year only over a third of all companies rated the practical use of this technology as irrelevant [13]. A significant increase has become apparent in recent years, with almost three quarters (i.e. 73%) of all companies surveyed already using cloud computing solutions [14]. The number of companies that consider cloud computing solutions to be irrelevant also decreased significantly during the observed period from 34 to 9%. According to a forecast by Gartner from 2018, around 80% of companies will completely abandon their local, traditional data centers by 2025, give up their own hardware capacities and transfer to external, centralized data centers [15, 16].

This development has resulted in a significant increase in the amount of energy required in data centers [2], which is expected to pose a major challenge for data centers [19]. In addition to the research project TEMPro (Total Energy Management for Professional Data Centers) [20], this paper will examine resource efficiency from the point of view of material efficiency, i.e. the relationship between the cost of materials for a product and the benefit resulting from product [21]. Therefore a joint consideration of durability, operating time and the usage along the product life cycle has to be made. One criteria to which all products in the IT and data center infrastructure are subject is obsolescence.

3 Obsolescence in Information and Communication Technology

Obsolescence is an unavoidable phenomenon [10] and thus affects all products in the data center infrastructure and IT infrastructure as well as licenses and services. Thus, in terms of sustainable development, organizations such as data centers have to face the issues of obsolescence.

The fact of its existence cannot be reduced to the aspect of products becoming unusable as natural obsolescence, but other reasons can be the cause of it. Accordingly, obsolescence can be divided into the types of material, functional, psychological and economic obsolescence [22, 23] (see also Table 1). It was found that, especially in the context of information and communication technology, there has been an increase in material and functional obsolescence in recent years [21]. This development corresponds to observations made by the project partner T-Systems at its data center in Biere. The findings show that the product lifecycle management of software vendors such as Microsoft with their operating system "Windows" has a direct impact on the product lifecycle management of data centers, for example by discontinuing support for software products due to missing patches. Although this

Table 1 Types of obsolescence

Obsolescence type	Definition
Material obsolescence	Defects due to poor performance of materials or components [23]
Functional obsolescence	New requirements are made on the functionality of a product, which are not fully met by the product itself. "An existing product becomes obsolete by the introduction of a new one that better fulfils its function" [26]
Psychological obsolescence	In this context, the targeted influence on the psyche of the consumer is meant [27]. A product is regarded as obsolete despite its proper quality, because it seems less desirable for fashion reasons or other changes [26]
Economic obsolescence	A defective product could be technically repaired, but the economic cost of repair is not worth it

is no proof of planned obsolescence,[3] but provides an interesting aspect for further research into this research area.

As shown in the study commissioned by the German Federal Environment Agency on the "Influence of the lifetime of products on their environmental impact" [22], both vendors and customers can influence the lifetime of products along the product life cycle, so that a statistical expectation arises as to how long a product can be used. In the context of obsolescence, this means that life expectancy and actual product life may differ [25].

In the context of sustainability, unpredictable obsolescence can be a significant cost factor especially if the requirements placed on an organization's processes cannot be fully met. This cost factor should be considered critical if additional supplies, spare parts or maintenance are difficult to obtain. From a social point of view, obsolescence-related downtime of a data center can have corresponding social impacts and challenges. The environmental impacts of increasing waste volumes, especially where these could be avoided through better planning, monitoring or reviews throughout the product life cycle, are also challenging.

Therefore, it is crucial for organizations to adopt a goal-oriented strategy to deal with obsolescence, minimizing the negative impact on sustainability goals. One way is adopting a management approach to obsolescence.

4 Obsolescence Management

4.1 Definition of Obsolescence Management

Research on obsolescence looks for ways in which it can be adequately managed, reduced and solved [28]. Obsolescence management can be understood as a continuous and regularly reviewed process—i. e. a "set of interrelated or interdependent activities that transform inputs into results" [29]—with a focus on continuous improvement. The main objectives are to ensure the quality of a sustainable management and to reduce the negative effects of obsolescence. Obsolescence management can be divided into four phases according to the PDCA cycle: plan, do, check and act.

Alternatively to the approach of establishing obsolescence management as an isolated management system, it can be integrated with other management systems to consolidate resources of related areas.

By planning proactively and carefully, obsolescence management can help to minimize the effects and costs of obsolescence[4] [30]. In order to achieve this main objective and based on the assumption that the product life of (IT-)products can

[3]Planned obsolescence is defined as "strategies and procedures of vendors and retailers to accelerate the purchase of new products by shortening usage cycles" [24].

[4]Accordingly, direct costs and indirect costs associated with making an organization more sustainable are regarded as costs.

be planned, obsolescence management includes the following tasks: Provision of substitution goods and of suitable spare parts, an overall reduction or avoidance of breakdowns (by identifying components that are endangered in terms of the process [31] through the early recognition of impending breakdowns) and the involvement of vendors and disposal companies by exerting influence along the supply chain.

4.2 Obsolescence Management Phases

Plan An obsolescence management plan should describe how activities to "identify and mitigate the effects of obsolescence" [10] are to be implemented. In this context, organization-specific causes, effects, risks, occurrence probabilities of obsolescence and possibilities for improvement and extension of the product life cycle are also identified and recorded. Depending on the market power of an organization, it may be useful to analyze and document influences on vendors along the supply chain and to plan in cooperation with them common obsolescence management objectives.

During planning, it should be determined which procedure is to be used to manage obsolescence for data center operations and for data center infrastructure products. Possibilities are:

- **Reactive Approach**: No measures are taken before obsolescence occurs, but only after it has occurred, for example, repairing obsolete products or procuring new products.
- **Proactive Approach**: Preventive measures are taken to achieve the objectives, whereby the risk classification and the predictable probability of obsolescence occurring are essential. Examples of measures are the proactive procurement of spare parts and substitute goods, the assurance of reparability as well as an analysis of the product design and installed components with a classification of critical components.
- **Strategic Approach**: Both above mentioned approaches are combined and it is determined for which products a reactive or proactive approach should be used.

Do The implementation of an obsolescence management should be obsolescence-resistant, i.e. it should take into account obsolescence in a permanent and consistent manner to steer towards the planned objectives of obsolescence management. In data center operations, this implies the obsolescence-resistant design of all data center infrastructure products along their life cycle as well as the materials, components and interfaces used in products.

In this context, the factor modularity, i.e. "the use of exchangeable units" [10], has to be taken into account. In the case of obsolescence, an individual part (or a subsystem) can be replaced without having to replace directly or indirectly related components at the same time. With regard to the use of software, it is recommended for an obsolescence-resistant design to possess the source code in order to be able to make appropriate maintenance decisions independently of external service providers. This can be achieved, for example, by using open source software. For proprietary software, it is crucial that the source code is available within the organization and can be maintained by the organization itself. An open system architecture that uses free standards, public protocols and common interfaces offers possibilities to replace obsolete modular software or hardware components and can help to reduce obsolescence costs in the medium and long term [10]. Furthermore, an analysis of future market developments is recommended in order to predict changes in the availability and sustainability of components and thus minimize the risk of obsolescence, e.g. through the procurement of long-lasting technologies and infrastructure products. Beyond the technologies, this also applies to the consideration of durable materials used in individual components [10].

Check It is important to make the performance resulting from the do-phase measurable with regard to the objectives defined in the plan-phase. Various evaluations of the effects of obsolescence in terms of sustainability are conceivable, such as the evaluation of environmental aspects, the identification of cost optimization potentials, and risk assessment based on obsolescence. In this context, the implementation of a performance measurement system is recommended. Particular attention should be paid to the comparability and continuity of key performance indicators over a certain period of time in order to be able to perform time series analyses with regard to the measurement and subsequent evaluation of the performance of obsolescence management. The consistent implementation of a performance measurement system also provides the opportunity for benchmarking [32].

Act This phase aims at an optimization with regard to the collected key performance indicators as well as an improvement of the measured performance. Measures are taken to continuously reduce obsolescence effects of products and services and to correct, reduce or prevent undesired effects.

5 Assessment of the Possibilities of Obsolescence Management in Data Centers

As already mentioned, a distinction can be made between IT infrastructure products, for which customers are responsible in traditional outsourcing, and data center infrastructure products, which the data center operators provide for their customers. Part of the latter are various large components, such as diesel generators and cooling technology, which are subject in their use and applicability to legal regulations and governmental authorizations. Due to changes in the legal situation, obsolescence regularly occurs in this area, so that, for example, transitional periods for the replacement of coolant have to be considered proactively in order to avoid negligently induced obsolescence.

For the IT infrastructure of data centers, several measures of obsolescence-resistant design have already been implemented in the past, including, for example, the consideration of modularity, whereby individual components instead of entire systems can be replaced in the case of a defect. For individual modules such as solid-state drive (SSD) hard disks or graphical processing unit (GPU) cores, it is possible to proactively determine states in the product life cycle of a hardware product at software level using permanent data transfer, so that strategic forecasts regarding durability can be made. Based on these forecasts, obsolescence management measures can be applied. Reasons for obsolescence tend to be functional at the software level, whereas reasons for obsolescence of hardware are manifold. Historically, products have increasingly become obsolete due to technical advances and energy efficiency criteria. The cyclical replacement of products is also considered as a cause of obsolescence as well as product changes and discontinuation. With regard to IT infrastructure, a shift in the liability limits towards the vendors and service providers responsible for installation, maintenance and replacement has been observed.

An essential component in large computer centers is thus the commissioning of service providers, also with the aim of preventing economic obsolescence by initiating a warranty claim. Due to this recent development, the obsolescence of services has become a significant issue. In the context of service management, orders are historically evaluated to identify anomalies, such as negative cost effects on a group of products.

All in all, there is an awareness of the challenges of obsolescence in large data centers and there are isolated approaches to obsolescence management according to DIN EN ISO 62402. In 2019, for example, a significant increase in the number of requests for sustainability and Green IT at the T-Systems data center in Biere was recorded. Data centers could meet this interest in sustainable IT infrastructures through certifications (e.g. in the form of sustainability or obsolescence certificates).

Due to the complexity and variety of products in IT and data center infrastructures, obsolescence is a cross-cutting challenge for different areas of the data center. Consequently, obsolescence has a direct impact on management tasks such as IT service management, product life cycle management (PLM), supply chain management (SCM), platform management and procurement management. These interdependencies must be taken into account and therefore, in terms of sustainability, approaches of obsolescence management should be integrated into existing management systems.

In the past, the high costs of reactive measures to deal with obsolescence were shifted to both manufacturers and external service providers, so that the sustainable use of products along their product life cycle could often not be adequately monitored. In order to counteract this disadvantage, possibilities of influencing the obsolescence-resistant process design of vendors and service providers must be identified and exploited. In this way, ecological potentials can be achieved in procurement processes by paying more attention to environmental certification.

Obsolescence in the IT infrastructure can be proactively detected by hardware monitoring and should also be used in local company-owned data centers. Obsolescence has a significant effect on services, since many service partners are commissioned in large data centers. Consequently, obsolescence must be identified in service management and patterns in the use of services must be recognized, for example, particularly frequently ordered maintenance services for a product or product group. Subsequently, conclusions should be drawn in strategic obsolescence management about the ordering of services and appropriate measures should be taken.

The challenge here is to involve (top-)management at all levels of obsolescence management and to define and consistently pursue an obsolescence management strategy. The involvement of all participants along the value chain in obsolescence management and the effective communication and cooperation in the complex computing center environment must be seen as a challenge, especially in internationally intertwined supply chains. Large data centers in particular now have considerable market power and can make a significant contribution to sustainable development through obsolescence management.

6 Conclusion and Future Work

This paper discussed the question of how data centers can advance sustainable development in dealing with the phenomenon of obsolescence. Up to now, many companies have transferred the responsibility for their IT infrastructures to external data centers due to liability reasons. Since the responsibility for a more sustainable design of information and communication technology along the product life cycle lies mainly with vendors and service providers, it is advisable for data center management to take advantage of their strong market power in procurement processes. An essential aspect in large data centers is the commissioning of service providers. This should be done with the aim of preventing economic obsolescence by initiating a warranty claim. Furthermore, the obsolescence of services is of great importance too. Within the context of service management, orders should be subsequently evaluated with the aim of identifying anomalies regarding the obsolescence of specific product groups.

Overall, there is an awareness of the challenges of obsolescence in large data centers and isolated approaches to obsolescence management according to DIN EN ISO 62402 can be found. Consequently, the importance of service management for data centers is increasing. This requires a proactive response to changes and cancellations of service providers and to avoid dependencies on vendors. For data center infrastructure products, changes in approvals and laws are mainly responsible for obsolescence. It is recommended to implement predictive obsolescence management approaches to prevent negligent obsolescence.

Basically, approaches of obsolescence management should be anchored in the management systems of the data centers in order to advance sustainable development across all organizational units. Due to the growing importance of data centers resulting from the increasing use of outsourcing solutions as well as the growth in capacity of large centralized data centers, issues of sustainable development, in particular criteria of resource efficiency, will become increasingly important in the future. Customers of data centers are increasingly demanding that IT solutions—as part of their own value creation—become more sustainable and that this is documented. A more sustainable approach to obsolescence could be developed through certifications beyond DIN EN ISO 62402.

References

1. Gabler Wirtschaftslexikon Online (2018). Stichwort „Rechenzentrum“, https://wirtschaftslexikon.gabler.de/definition/rechenzentrum-42319/version-265670. Last accessed 27 Apr 2020
2. Hintemann, R.: Nachhaltige Rechenzentren: Viel mehr als nur Energieeffizienz (2018). https://www.funkschau.de/datacenter/artikel/159905/. Last accessed 18 Oct 2019
3. Marx Gómez, J., Gizli, V.: Ein Framework zur ganzheitlichen Steigerung der Energieeffizienz in Rechenzentren durch eine Konformitätsprüfung zertifizierter Kennzahlen. In: Arndt H.-K., Marx Gómez, J., Wohlgemuth, V., Lehmann, S., Pleshkanovska, R. (eds.) Nachhaltige betriebliche Umweltinformationssysteme: Konferenzband zu den 9. BUIS-Tagen, pp. 3–8. Springer Gabler, Wiesbaden (2018)
4. Carl von Ossietzky Universität Oldenburg (2017). Der wahre Energieverbrauch von Rechenzentren: Projekt „TEMPro“ nimmt gesamten Lebenszyklus der Infrastruktur in den Blick. https://www.presse.uni-oldenburg.de/mit/2017/030.html. Last accessed 11 Nov 2019
5. Gabler Wirtschaftslexikon Online (2018). Stichwort „Green IT“, https://wirtschaftslexikon.gabler.de/definition/green-it-53166/version-276261. Last accessed 27 Apr 2020
6. Urban, T., Arndt, H.-K.: Das Common Information Model als Datenmodell zur Bestimmung von Green-IT-Kennzahlen der Lebenszyklusphase Make von IT- Dienstleistungen. In: Arndt H.-K., Marx Gómez, J., Wohlgemuth, V., Lehmann, S., Pleshkanovska, R. (eds.) Nachhaltige betriebliche Umweltinformationssysteme: Konferenzband zu den 9. BUIS-Tagen, pp. 297–303. Springer Gabler, Wiesbaden (2018)
7. Schwab, W.: Experton Group (2009): Green IT in der Wirtschaftskrise, das optimale Rechenzentrum. https://www.channelpartner.de/green-it/275141/. Last accessed 29 Feb 2020
8. Duden (2019). Obsoleszenz. https://www.duden.de/suchen/dudenonline/Obsoleszenz. Last accessed 10 Aug 2019
9. Deutsches Institut für Normung (DIN) e.V. (eds.): DIN EN ISO 14001 Umweltmanagementsysteme - Anforderungen mit Anleitung zur Anwendung (ISO 14001:2015), DIN-Normenausschuss Grundlagen des Umweltschutzes (NAGUS). Beuth Verlag, Berlin/Wien/Zürich (2015)
10. Deutsches Institut für Normung (DIN) e.V. (eds.). DIN EN 62402:2017–09 – Entwurf: Obsoleszenzmanagement (IEC 56/1716/CD:2016). DKE Deutsche Kommission Elektrotechnik Elektronik Informationstechnik in DIN und VDE. Beuth Verlag, Berlin/Wien/Zürich (2017)
11. Bundesverband Informationswirtschaft, Telekommunikation und neue Medien e. V. (eds.): Betriebssicheres Rechenzentrum – Leitfaden. Berlin (2013)
12. T-Systems (2020). Rechenzentrum Biere: Deutschlands digitales Herz - Im Bördeland bei Magdeburg betreibt T-Systems eines der größten Cloud Data Center Europas. https://www.t-systems.com/de/de/ueber-t-systems/unternehmen/innovation-management/rechenzentrum-biere. Last accessed 27 Apr 2020
13. KPMG, Bitkom Research (2016). Cloud Monitor 2016: Cloud-Computing in Deutschland - Status quo Perspektiven. https://hub.kpmg.de/cloud-monitor-2016. Last accessed 11 Nov 2019

14. KPMG, Bitkom Research (2019). Cloud Monitor 2019: Public Clouds und Sicherheit im Fokus. https://hub.kpmg.de/cloud-monitor-2019. Last accessed 11 Nov 2019
15. Costello, K.: Gartner Top 10 Trends Impacting Infrastructure & Operations for 2019 (2018). https://www.gartner.com/smarterwithgartner/top-10-trends-impacting-infrastructure-and-operations-for-2019/. Last accessed 15 Jan 2020
16. Ostler, U.: Datacenter Insider - Gartner prognostiziert: Datacenter- Trends 2019 und darüber hinaus (2018). https://www.datacenter-insider.de/datacenter-trends-2019-und-darueber-hinaus-a-785604/. Last accessed 15 Jan 2020
17. KPMG, Bitkom Research (2017). Cloud Monitor 2017: Cyber Security im Fokus. https://hub.kpmg.de/cloud-monitor-2017. Last accessed 11 Nov 2019
18. KPMG, Bitkom Research (2018). Cloud Monitor 2018: Strategien für eine zukunftsorientierte Cloud Security und Cloud Compliance. https://hub.kpmg.de/cloud-monitor-2018. Last accessed 11 Nov 2019
19. Hintemann, R.: Wachstumsschub durch Cloud Computing – Effizienzgewinne reichen nicht aus: Energiebedarf der Rechenzentren steigt weiter deutlich an (2020). https://www.borderstep.de/wp-content/uploads/2020/03/Borderstep-Rechenzentren-2018-20200327rev.pdf. Last accessed 24 Apr 2020
20. Hintemann, R.: Total Energy Management for professional data centers (Tempro) Ganzheitliches Energiemanagement in professionellen Rechenzentren (2019). https://www.borderstep.de/projekte/total-energy-management-for-professional-data-centers/. Last accessed 24 Apr 2020
21. Bundeszentrale für politische Bildung (2014). Materialeffizienz. https://www.bpb.de/gesellschaft/umwelt/klimawandel/180054/materialeffizienz. Last accessed 03 Apr 2020
22. Prakash, S., Stamminger, R., Dehoust, G., Gsell, M., Schleicher, T., Gensch, C.-O., Graulich, K.: Einfluss der Nutzungsdauer von Produkten auf ihre Umweltwirkung: Schaffung einer Informationsgrundlage und Entwicklung von Strategien gegen ‚Obsoleszenz': Abschlussbericht (2016). https://www.umweltbundesamt.de/sites/default/files/medien/378/publikationen/texte_11_2016_einfluss_der_nutzungsdauer_von_produkten_obsoleszenz.pdf. Last accessed 11 Dec 2019
23. Oehme, I., Jacob, A., Cerny, L., Fabian, M., Golde, M., Krause, S., Löwe, C., Unnersta, H.: Strategien gegen Obsoleszenz: Sicherung einer Produktmindestlebensdauer sowie Verbesserung der Produktnutzungsdauer und der Verbraucherinformation (2017). https://www.umweltbundesamt.de/sites/default/files/medien/1410/publikationen/2017_11_17_uba_position_obsoleszenz_dt_bf.pdf. Last accessed 11 Dc 2019
24. Schridde, S.: Die Dimensionen der geplanten Obsoleszenz - Öffentliche Anhörung des Parlamentarischen Ausschuss für nachhaltige Entwicklung am 17.12.2015 Weiterentwicklung der Produktverantwortung (2015). https://www.murks-nein-danke.de/blog/download/A-Drs.%2018(23)25-4.pdf. Last accessed 23 Dec 2019
25. Poppe, E., Longmuß, J.: Geplante Obsoleszenz: Hinter den Kulissen der Produktentwicklung. Transcript Verlag, Bielefeld (2019)
26. Packard, V., McKibben, B.: Waste Makers. Lg Publishing, Brooklyn (2011)
27. Reuß, J., Dannoritzer, C.: Kaufen für die Müllhalde: Das Prinzip der geplanten Obsoleszenz. Orange-press, Freiburg (2013)
28. Romero Rojo, F.J., Roy, R., Shehab, E.: Obsolescence management for long-life contracts: state of the art and future trends. Int. J. Adv. Manuf. Technol. **49**, 1235–1250 (2009)
29. Deutsches Institut für Normung (DIN)) e.V. (eds.). DIN EN ISO 50001:2018–12 Energiemanagementsysteme - Anforderungen mit Anleitung zur Anwendung (ISO 50001:2018), DIN-Normenausschuss Grundlagen des Umweltschutzes (NAGUS). Beuth Verlag, Berlin/Wien/Zürich (2018)
30. Schnieder, L.: Obsoleszenzmanagement. In: Schnieder, L. (ed.) Strategisches Management von Fahrzeugflotten im öffentlichen Personenverkehr, pp. 221–239. Springer Verlag, Berlin/Heidelberg/New York (2018)

31. Wiesböck, J.: Materialbeschaffung: Obsoleszenz-Management – der Schlüssel zur Langzeitverfügbarkeit (2017). https://www.meilensteine-der-elektronik.de/obsoleszenz-management-der-schluessel-zur-langzeitverfuegbarkeit-a-576901/. Last accessed 24 Jan 2020
32. Clausen, J., Kottmann, H.: Umweltkennzahlen im Einsatz für das Benchmarking. In: Seidel, E. (eds.) Betriebliches Umweltmanagement im 21. Jahrhundert, pp. 255–265. Springer Verlag, Berlin/Heidelberg/New York (1999)

The Eco-label Blue Angel for Software—Development and Components

Stefan Naumann, Achim Guldner, and Eva Kern

Abstract Energy consumption induced through information technology, i.e. hardware and software, is constantly increasing. In this article, we present the "Blue Angel", a label that evaluates and classifies the resource and energy efficiency of software. In particular, the process by which the label was developed is presented. We also describe the components of the Blue Angel: Energy and resource efficiency of the software product, hardware useful life and user autonomy. Finally, we give an outlook on the possibilities for expansion, since the first version of the label focuses especially on desktop software, mainly because of the complex measurement issues. According to evaluations from expert workshops and interviews, this is the best starting point for reliable results and certifications.

Keywords Green software · Blue angel · Energy efficiency

1 Introduction

The increasing digitization is realized through information and communication technologies (ICT), which consume considerable amounts of energy. In the public debate, negative effects of digitization on the environment are associated with the energy consumption of hardware. However, many people lack awareness of the environmental impact of software. Software products are immaterial, are not subject to wear and tear, and no waste is produced when they are disposed of. Viewed in isolation, software appears to be a fully sustainable product. In practice, however, software products with the same or similar functions can differ significantly in their impact

S. Naumann (✉) · A. Guldner · E. Kern
Environmental Campus Birkenfeld, Institute for Software Systems, Trier University of Applied Sciences, P.O. Box 1380, 55761 Birkenfeld, Germany
e-mail: s.naumann@umwelt-campus.de

A. Guldner
e-mail: a.guldner@umwelt-campus.de

E. Kern
e-mail: e.kern@umwelt-campus.de

A. Kamilaris et al. (eds.), *Advances and New Trends in Environmental Informatics*, Progress in IS, https://doi.org/10.1007/978-3-030-61969-5_6

on natural resources [1]. Accordingly, environmental impacts can be caused by software, as software induces the hardware energy consumption. Software has a special role to play, because its properties and functions determine what hardware capacities are required. Thus, the design of software products has an influence on the energy requirements of information and communication technology.

In order to draw attention to the role of software in the ICT sector and at the same time provide orientation for users, procurers and developers, the Environmental Campus Birkenfeld, together with den German Oeko-Institut developed a Blue Angel[1] for energy and resource efficient software products on behalf of the German Federal Environment Agency. This paper presents relevant aspects of the development of the eco-label as well as the award criteria and approaches behind it. It addresses the question of how scientific results (catalogue of criteria for sustainable software products including developed verification methods) can be transferred into practical application.

2 Background

2.1 Eco-label Blue Angel

The "Blue Angel" eco-label is an important component of Germany's product-related environmental policy: The label serves not only as an orientation for consumers, who can use it to identify leading ecological products, but also for public procurement, which can use the award criteria as a basis for tenders. After the EU Directive 2014/24/EU has been transposed into national law, it will even be possible to demand products and services in public tenders to bear the eco-label or otherwise demonstrate compliance with the criteria [2]. For manufacturers and dealers, the eco-label offers the opportunity to label their environmentally friendly products to gain a market advantage. In addition, the eco-label criteria provide manufacturers with the technical parameters for optimizing their own products and aligning product development with these benchmarks. At European level, the Blue Angel award criteria are incorporated into the eco-design process and the further development of the EU eco-label. In total, about 12,000 products and services in about 120 product categories currently bear the Blue Angel.[2]

In this research project, software products were examined as a new product group for the Blue Angel and criteria for the eco-label were developed. Due to increasing digitization, the relevance of energy and resource efficiency of software products is also increasing, as is the general market for software products.

[1] https://www.blauer-engel.de/en/get/productcategory/171, last accessed 2020-04-30.

[2] blauer-engel.de, last accessed 2020-04-30.

For example, sales of software in Germany rose from 14.3 billion Euros in 2007 to 26 billion in 2019, and an increase to 27.6 billion is forecast for 2020 [3, 4]. The software segment is the fastest-growing part of the ICT market, for which a sales growth of 1.5% is expected. While sales in consumer electronics are declining, investments in telecommunications services are increasing. In addition to the sales increases in the ICT segment, the energy consumption of telecommunications networks and data centers is also rising, likely from 18 TWh in 2015 to 25 TWh in 2025 [5]. In addition, data traffic is increasing, especially with mobile devices: here, the data volumes transmitted via the Internet rose by more than 800% between 2011 and 2016 to 913 million GB in 2016. Meanwhile, in the fixed network the data volume increased 4.5 times over the same period [6].

The aim of the newly developed eco-label is to reduce the overall energy consumption of ICT and to increase the efficiency of resources. Products whose manufacturers disclose information about their products are particularly highlighted by the Blue Angel for this transparency. In addition, a product whose manufacturer is actively committed to improving the resource and energy efficiency of its software products is also labeled.

2.2 *Other Labels*

There are no other eco-labels known for software products. In the following, eco-labels are listed which refer to the hardware sector with a focus on Green IT:

- Blue Angel for other products (e.g. computer centers, computers, car sharing)
- European eco-label, so-called "Euro Flower" (e.g. notebooks, computers, desktop computers)
- ENERGY STAR (e.g. computers, monitors, printers, servers)
- TCO labels (e.g. flat screens, desktop and notebook PCs, printers, mobile phones)
- ECMA-370 (The Eco Declaration, TED, international standard for product declarations)
- EPEAT (global rating system for green(ere) electronics, e.g. PCs, monitors).

3 Progress of the Investigation

The overall objective of the project was to develop the basic award criteria [1] for the Blue Angel eco-label and to establish a Blue Angel for software products.

Due to the nature of the object of investigation within the product group "Resource-Efficient Software", the implementation procedure had to be altered from a "classical" approach towards a Blue Angel for physical product groups like PROSA [7]. The work package was divided into the following tasks (Table 1).

In the past, questions regarding the resource efficiency of software could not be answered sufficiently and there was no uniform system in place. Thus, available

Table 1 Tasks and time schedule of the research project to develop a Blue Angel for software

Task
Task 1: Delimitation of suitable software product groups
Task 2: Field tests
Task 3: Practical workshop
Task 4: Draft version of the "Basic Award Criteria"
Task 5: Expert hearing
Task 6: Jury conference
Task 7: Documentation

market data and practical benchmarks were available only in a very limited fashion when developing the Basic Award Criteria for a Blue Angel. Therefore, *Task 1* focused on the identification of suitable software types within the course of a market analysis, the development of standard usage scenarios, energy measurement can be laid on, and the definition of minimum requirements, describing resource-efficient software products. In addition, a survey was conducted to investigate which software products organizations like companies, data processing service centers and public institutions typically commissioned or procured.

Due to the complexity of "software products" as the label's object, it first had to be narrowed down. Therefore, the award criteria and procedures were developed, based upon the identified, suitable software product groups. This finally resulted (in Task 4) in a limitation of the scope of the first version of the Blue Angel to application software that has a user interface and can be run on a defined desktop reference system. Even if the criteria can also be applied to other systems (like client–server systems), the scope was limited, because reliable results can be gained especially for desktop systems. This results from the complex measurement issues of distributed systems (client, network, server), where it is possible to measure/estimate the consumption in different ways. However, for a formal certification it is necessary to be more concrete. Thus, the Federal Environment Agency decided to start with desktop systems. An extension is planned in 2021.

In *Task 2*, companies were identified and contacted that both develop or produce software products with the appropriate requirements and, at the same time, were generally interested in an application for the label. This procedure (see Fig. 1) made it possible to find industry partners who evaluated the proposed evaluation criteria in field tests.

At first, the partners were informed in detail about the proposed criteria and their applicability. They then applied the criteria to their product with the help of authors' institute staff. The planned measurement and recording procedures, the award criteria and their indicators, as well as the advantages resulting from the use of the label were presented and examined in detail. In particular, this included recording the standard usage scenarios that are required for the award criteria. The scenarios were developed in conjunction with the companies for their respective software products.

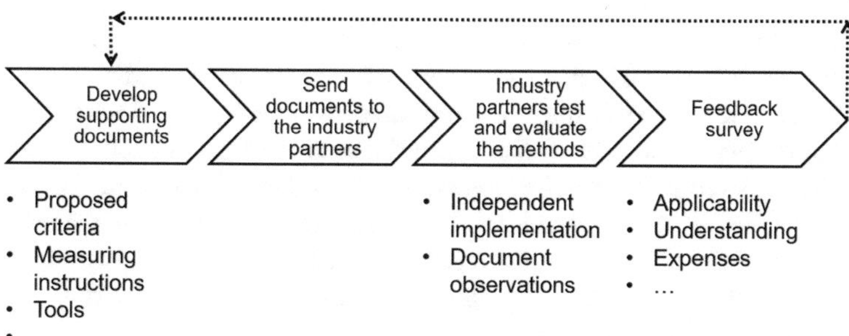

Fig. 1 Process of the field test to evaluate the proposed criteria together with industry partners

In *Task 3*, the results obtained through the field tests and prior research were compiled, discussed, and elaborated in a practical workshop. Here, the results from the companies participating in the field test were presented. The experience and points of criticism obtained during the field test were integrated into a new version of the Basic Award Criteria and the procedure for recording and developing standard usage scenarios. This, as well as the obligation of applicants to release the necessary information for the compliance verification, was presented and discussed.

In this process, the issue of the standard usage scenarios was given much attention. This is because they represent the typical usage of the software product and are required by multiple potential award criteria. Several strategies were discussed, to eliminate preferring or penalizing any software products to be compared (in case of software products with a comparable range of functions). For the first version of the Blue Angel, it was decided that the applicants would create, document, and publish the standard usage scenario themselves, using a template, developed during the research project. The scenario used to verify the compliance must include those functionalities typically used in the software product. It then must be submitted with the application documents and is checked for plausibility by auditors.

Based on the results from Tasks 1 to 3, the draft version of the Basic Award Criteria was continuously developed and validated in *Task 4*. In particular, this includes the definition of the compliance verification to be provided for all criteria and their indicators along the software properties resource and energy efficiency, potential hardware operating life and user autonomy. Due to the special nature of the label's object, the Basic Award Criteria were extended to include additional information on the procedure for the compliance verification of criteria "hardware utilization" and "electrical power consumption in idle mode/when a standard use scenario is executed", see [1]. This includes measuring instructions for these criteria, the definition of a data format for providing product information on resource and energy efficiency, and information on reference systems and standard usage scenarios. These guidelines have been formalized and included as appendices in the award criteria.

After a consultation with the German Oeko-Institut and the Federal Environment Agency, in *Task 5*, the draft for the Basic Award Criteria was handed over to the RAL Institute (the organization which maintains the Blue Angel) for further steps.

In an expert hearing, scheduled by RAL on October 23rd, 2019, all criteria and indicators were discussed with experts from software production, trade associations, authorities, and research institutes. The proposed changes and decisions that emerged during the discussion were recorded and influenced the final version of the criteria catalog. The resulting final version, which was agreed with the Federal Environment Agency, was finally made available to RAL in *Task 6* and presented at the meeting of the Environmental Label Jury on 11 December 2019. Following the positive decision of the jury, the final version including all appendices was produced and published on the Blue Angel website in January 2020.

Task 7 consisted of producing the documentation needed for utilization of the work.

4 Stakeholders Within the Eco-label Progress

The Blue Angel for energy- and resource-efficient software products pursues various goals: on the one hand, it aims to make software companies and thus software developers aware of the environmental impact of software. On the other hand, the eco-label provides orientation especially for purchasers and consumers. Therefore, to get substantial criteria and to spread the results within the community it is necessary to integrate several stakeholder groups.

4.1 Perspective of Purchasers

In order to find out which software products are typically procured, and which are ordered, a survey of procurers was conducted. A questionnaire for a written survey as well as a guideline for interviews was created. Contents of the survey were:

- What is the typical procedure for your software procurement?
- Do environmental criteria play a role in your procurement process?
- What are your evaluation criteria for software?
- How do you rate labels (especially environmental and energy) in the procurement process?

The structured interviews were conducted at two universities and two public authorities between October 2018 and May 2019. The results show that software is predominantly procured, not commissioned. Price and functionality are primary criteria for a purchase. Asked about criteria that are of interest in software procurement, the following aspects were mentioned: Required resources, power consumption, costs, environmental friendliness, information on the test phase, functionality, planned further development of the products, license model, stability of the products, existing backdoors (e.g. IT security).

The idea of an eco-label for software products was welcomed in principle in order to include the award criteria contained therein in the procurement process where appropriate. In particular, however, the selection of procured software products depends particularly on their functionality and their suitability for the requirements of future users.

4.2 Perspective of Software Companies

During the development of the Basic Award Criteria, all interested software companies were welcomed to contribute, since one of the objectives of the Blue Angel is to raise awareness of the topic “environmental impacts of software".

In order to be able to assess the criteria and minimum requirements from a practical perspective, e.g. a field test was conducted (see Sect. 3). In this way, the expertise of the practice partners could be included in the further development of the award criteria. Among other things, the handling of updates, for example the proof for the criterion “downward compatibility” and the criteria regarding product information were revised. A practical workshop was held in June 2019 in order to establish a direct exchange with the software companies. Based on the feedback, the criteria were revised again and then discussed with experts in a hearing (see Sect. 3). The results lead to the definition of the scope of the eco-label for application software with user interface and helped sharpen individual criteria.

4.3 Perspective of the Users of Software

In order to also include the view of software users on an eco-label for software products in the development process of the award criteria, their interest in environmental issues of software was analyzed. For this purpose, an online survey was conducted in August/September of 2016 as part of a dissertation, in which the questions were primarily directed at German software users. The total sample comprised n = 854 participants. The incomplete data sets were excluded in the follow-up work. A total of 712 questionnaires were answered completely and used for data analysis. 55.8% of the respondents were female (40.7% male, 3.5% not specified). About one third of the participants were between 20 and 39 years old (20 to 29 years: 39.3%, 30 to 39 years: 29.4%). Most of the participants see themselves primarily in the role of users (78.2%). Regarding personal IT skills, they consider themselves to be competent players (40.4%) and advanced beginners (36.5%). Further details on the survey can be found in [1, 8].

5 Main Components of the Blue Angel for Software

5.1 Resource and Energy Efficiency

Depending on the characteristics and configuration of the software product, the fulfillment of a certain functionality can lead to different demands on hardware resources and thus to different degrees of environmental impact. The functionality is defined by a method called "standard usage scenario", which describes the typical usage of a software with a user interface. The overall aim of the Blue Angel is to award software products with the highest possible energy and resource efficiency, i.e. to provide the functionality with the lowest possible energy and resource consumption. Therefore, the following aspects must be recorded:

- Required minimum system requirements
- Hardware utilization and electrical power consumption in idle
- Hardware usage and energy requirements when running a standard usage scenario
- Support of the energy management.

5.2 Potential Hardware Lifetime

Due to the growing hardware-demands of software, hardware must increasingly be replaced, i.e. the hardware service life is shortened. The question arises how the renewal cycles of hardware can be decoupled from those of software. The longest possible useful life should be aimed for in order to minimize the resources required for the provision of new hardware. Therefore, with the criterion of Downward compatibility the software producer must confirm that the software product to be labelled is executable on a reference system that is available at least since five years at the application date.

Former versions of the criteria collection [1] included a criterion called "platform independence and portability" (Can the software product be operated on various currently widespread productive system environments (hardware and software) and can users switch between them without disadvantage?) and "hardware efficiency" (Does the amount of hardware capacity used remain constant over time as the software product is further developed, even if functions are expanded?). These have been replaced in the final Blue Angel award criteria, because on the one hand the eco-label for a software product is applied for on a specific platform (reference system). On the other hand, values over a longer period of time would be necessary to demonstrate hardware efficiency. Thus, these aspects were restated in the award criteria "requirements during the contract period", which stipulate that the software must remain executable on the same reference system and that the energy requirement may not increase by more than 10% compared to the values at the time of application.

5.3 Autonomy of Use

If it is assumed that many users are interested in a resource-saving usage of software, they should be provided with a basis for decision-making. Therefore, they require insights into the product, the means to reduce the capacities used by a software product, and corresponding information. The options should be available at such a low threshold that even users who are not familiar with the technology can act autonomously. The following aspects are to be demonstrated or made available for this purpose:

- Data formats
- Transparency of the software product
- Continuity of the software product
- Uninstallability
- Offline capability
- Modularity
- Free of ads
- Documentation of the software product, the license, and usage conditions

In addition, the former catalog of criteria also mentioned the aspects of "transparency of process management" (Does the software product make users aware that it automatically starts or continues processes in the background that may not be used?), "ability to delete generated data" (Are users sufficiently supported in deleting data generated during the operation of the software product that they have not explicitly created?) and "maintenance functions" (Does the software product offer easy-to-use functions that enable them to repair damage to data and programs that has occurred?). These have been removed from the award criteria due to complex traceability. The aspects "modularity" and "free of ads" were added after the practical workshop due to feedback of the participants.

A more detailed derivation of individual criteria and their operationalization can be found in the final report "Development and application of evaluation bases for resource-efficient software taking into account existing methodology".[3]

6 Conclusion and Outlook

"What you can't measure you can't manage" is one main credos of life cycle analysis. In this paper, we described the development of the Blue Angel for software. This eco-label helps to classify and label software products which are more resource and energy efficient than other products. In the first step, transparency is necessary. Therefore, it is in the focus of the first iteration of the Blue Angel for energy and

[3] https://www.umweltbundesamt.de/publikationen/entwicklung-anwendung-von-bewertungsgrundlagen-fuer (available in German).

resource efficient software products. It provides tools for measurement, and assessment and the means to make the data on environmental impacts of software public. In the next steps two directions are interesting: to extend the Blue Angel to also include client–server systems and other software architectures, which are more complex to measure, and also to collect several measurement results in order to compare and to rank the resource efficiency in software groups of similar functionalities. Additionally, attempts to assess the environmental impact can be extended to more complex software products, like machine learning, where first approaches exist, like [9] or [10]. Furthermore, the extension to more demanding environments like embedded systems e.g. [11] or [12] or mobile devices e.g. [13] or [14] should be investigated, as these devices become more and more prominent, when compared to standard desktop computers. The overarching research question behind these is: How can environmental impacts of ICT be recorded, proven, and evaluated?

Acknowledgements This work was supported by the German Environment Agency under project number 3718 37 316 0.

References

1. Kern, E., Hilty, L.M., Guldner, A., Maksimov, Y.V., Filler, A., Gröger, J., Naumann, S.: Sustainable software products—towards assessment criteria for resource and energy efficiency. Fut. Gener. Comput. Syst. **86**, 199–210 (2018)
2. European Parliament: Directive 2014/24/EU of the European Parliament and of the Council of 26 February 2014 on public procurement and repealing Directive 2004/18/EC. https://eur-lex.europa.eu/legal-content/EN/TXT/?uri=celex:32014L0024. Last accessed 21 Apr 2020
3. Tenzer, F.: Umsatz mit Software in Deutschland in den Jahren 2007 bis 2020. https://de.statista.com/statistik/daten/studie/189894/umfrage/marktvolumen-im-bereich-software-in-deutschland-seit-2007/. Last accessed 23 Mar 2020
4. Meyer, M.: ITK-Märkte. https://www.bitkom.org/Marktdaten/ITK-Konjunktur/ITK-Markt-Deutschland.html. Last accessed 23 Mar 2020
5. Stobbe, L., Proske, M., Zedel, H., Hintemann, R., Clausen, J., Beucker, S.: Entwicklung des IKT-bedingten Strombedarfs in Deutschland. https://www.bmwi.de/Redaktion/DE/Downloads/E/entwicklung-des-ikt-bedingten-strombedarfs-in-deutschland-abschlussbericht.pdf. Last accessed 23 Mar 2020
6. Federal Network Agency for Electricity (2017)
7. Grießhammer, R., Buchert M., Gensch, C.-O., Hochfeld, C., Manhart, A., Reisch, L., Rüdenauer, I.: PROSA – Product Sustainability Assessment. Guideline. https://www.prosa.org/fileadmin/user_upload/pdf/leitfaden_eng_final_310507.pdf, 2007. Last accessed 30 Apr 2020
8. Kern, E., Guldner, A., Naumann, S.: Bewertung der Nachhaltigkeit von Software - Entwicklung einer Umweltkennzeichnung. In: Arndt, H.-K., Marx Gómez, J., Wohlgemuth, V., Lehmann, S., Pleshkanovska, R. (eds.) Nachhaltige Betriebliche Umweltinformationssysteme - Research - Konferenzband zu den 9. BUIS-Tagen, Springer Gabler, Wiesbaden (2018)
9. Henderson, P., Hu, J., Romoff, J., Brunskill, E., Jurafsky, D., Pineau, J.: Towards the systematic reporting of the energy and carbon footprints of machine learning (2020)
10. Strubell, E., Ganesh, A., McCallum, A.: Energy and policy considerations for deep learn-ing in NLP (2019)

11. Cherupalli, H, Duwe, H., Ye, W., Kumar, R., Sartori, J.: Determining application-specific peak power and energy requirements for ultra-low-power processors. ACM Trans. Compt. Syst. **35**(3), (2017)
12. Georgiou, K., Xavier-de Souza, S., Eder, K.: The IoT energy challenge: A software perspective. IEEE Embedded Syst. Lett. **10**, 53–56 (2018)
13. Bunse, C.: On the impact of code obfuscation to software energy consumption. In: Otjacques, B., Hitzelberger, P., Naumann, S., Wohlgemuth, V. (eds.) From Science to Society—New Trends in Environmental Informatics. Progress in IS, pp. 239–249. Springer International Publishing (2018)
14. Palomba, F., Di Nucci, D., Panichella, A., Zaidman, A., De Lucia, A.: On the impact of code smells on the energy consumption of mobile applications. Inf. Soft-Ware Technol. **105**, 43–55 (2019)

Sustainable Processes on the Last Mile—Case Study Within the Project 'NaCl'

Uta Kühne, Mattes Leibenath, Camille Rau, Richard Schulte, Lars Wöltjen, Benjamin Wagner vom Berg, Kristian Schopka, and Lars Krüger

Abstract Today the proportion of customers who prefer buying online increases in many product categories. The rising proportion of online orders and the associated returns have a significant effect on the existing logistics system. The city centers in particular are severely affected by the resulting traffic on the last mile with emissions, noise and traffic jams. In addition to these effects, the increasing pressure on performance and costs has a negative impact on the drivers employed by logistics service providers. The last mile, however, offers great potential for sustainable logistics processes. This article uses the sustainable crowd logistics project NaCl to show possible processes for last mile logistics.

Keywords Sustainability · SusCRM · Last mile logistics · Logistics process · Crowd logistics · Bundling

---

U. Kühne (✉) · M. Leibenath · C. Rau · R. Schulte · L. Wöltjen · B. W. vom Berg
An der Karlstadt 8, 27568 Hochschule Bremerhaven, Germany
e-mail: uta.kuehne@hs-bremerhaven.de

M. Leibenath
e-mail: mattesleibenath@gmx.de

C. Rau
e-mail: camille.rau@web.de

R. Schulte
e-mail: richard_schulte@outlook.de

L. Wöltjen
e-mail: lars@woeltjen.info

B. W. vom Berg
e-mail: benjamin.wagnervomberg@hs-bremerhaven.de

K. Schopka
Rytle GmbH, Schwachhauser Ring 78, 28209 Bremen, Germany
e-mail: kristian.schopka@rytle.com

L. Krüger
Weser Eilboten GmbH, Am Grollhamm 4, 27574 Bremerhaven, Germany
e-mail: Lars.Krueger@druckzentrum-nordsee.de

A. Kamilaris et al. (eds.), *Advances and New Trends in Environmental Informatics*, Progress in IS, https://doi.org/10.1007/978-3-030-61969-5_7

1 Introduction

The following definition by Gevaers et al. reflects the todays situation in urban logistics: 'The last mile is currently regarded as one of the more expensive, least efficient and most polluting sections of the entire supply chain' [1]. Within the supply chain management, the last mile is understood to be the overcoming from the transport hub to the final destination. At this part of the supply chain the customers experience directly the aspects of service provided by the courier, express and parcel (CEP) service provider. Therefor the logistics companies are in charge of satisfying the customers' demands to stay competitive.

An innovative approach for overcoming the last mile are on-demand platforms which are providing effective solutions for modern urban logistics. As the traffic worldwide is increasing rapidly [2] city logistics is facing various challenges on the economical, ecological and social level.

The stationary retail is losing more and more market shares to the e-commerce but as it has the advantage of customer proximity. This potential should be used by taking it into account in urban logistics.

While great challenges can be found within the last mile logistics, a lot of chances can be identified for the logistics companies. To face the challenges and benefit from the chances, the actual situation of the CEP-sector has to be analyzed with focus on different aspects and innovative solutions have to be developed. To follow a sustainable approach, it is important to examine solutions under the already mentioned aspects of ecology, economy and society and to combine them in one solution system while at the same time it has to be ensured that the requirements of the customers and furthermore, those of a sustainable last mile logistics are fulfilled.

2 Fundamentals and Innovations

2.1 State of the Art and Challenges

Nowadays it can be observed that consumers tend to buy products online. Studies from 2017, however, show that German consumer still prefer buying most products in the stationary retail, but at the same time these studies set out that a rapid change in the buying behavior in the direction of online trading can be seen. The proportion of customers who preferred buying stationary decreased in all product categories examined between 2016 and 2017, for example, the category electronics and computers lost 10% while the leading stationary shopping channel foods lost 5% [3].

The choice of shopping channel depends on the type of product, but it was forecasted that in 2020 every fifth purchase will be made online [4]. Many returns are associated with many orders and demand logistics to develop innovative approaches.

These developments lead to big challenges for logistics, especially on the last mile. In the year 2021 it is expected, that there will be a transport volume of 4.15

billion packages in the German market, which is meant to be handled by the already overstrained CEP-sector [5]. The increasing traffic load resulting from the rising number of shipments leads to traffic jams and environmental pollution, especially within the inner-city traffic. The last mile causes friction, because it is not only an obstacle to traffic but also has a negative impact on the quality of life and air. With 18.2 percent (166 Mio. T CO_2-equivalent) in 2016 traffic is the third largest cause of emissions in Germany [6] and has a huge negative impact on the environment.

There is great potential, especially on the last mile, to make supply chains more environmentally friendly in order to meet the general demand for sustainability in the economy and society and to improve the image of the industry. The use of electromobility instead of fuel-operated vehicles has positive effects on emissions, as it is locally emission-free. While the EU's climate protection targets call for an emissions reduction of 40 percent by 2030 [7], at the same time, surveys show that for example consumers like to be more environmentally friendly and ethical through the economic acting of brands [8].

Just like consumer behavior, this shows that incentives have to be created so that consumers can and are willing to participate in a more sustainable delivery process. Providers of logistics services have to act in order to meet the demands of the consumers as those are the ones that define the requirements. It must therefore be ensured that the supply chain is designed on the basis of the endpoints (the requirements of the customers). According to studies from 2017, the most important delivery criteria are free shipping, the possibility of tracking the shipment and satisfaction with the delivery service [9].

On the one hand the demand of public institutions and the customers for more sustainability are driving the industry to act more environmentally friendly, but on the other hand the consumers ask for free shipment and great service. Solutions have to be found to make the partly contrary goals compatible. The use of bundling, e.g., has the potential to realize the demand for cheap and at the same time sustainable delivery.

While sustainability is discussed on the level of ecology, it also has to be focused on the social aspects of sustainability. Several start-ups and companies today offer various solutions to meet the demand for flexible and cheap deliveries but at the same time those approaches often are missing social aspects such as fair payment, job stability and health insurance. Furthermore, these approaches have a destructive effect on the existing logistics labor market, since they often follow a crowd logistics approach that results in disruptive consequences for the logistics industry. Uber freight, for example, is a freight-on-demand service which takes over the brokerage part of the freight forwarder and in this way has the potential to make big changes in the traditional business model by replacing the freight forwarders with data and algorithms. As no human interaction would be needed anymore in this process design, it doesn't meet the requirements of social sustainability within the last mile. Moreover, this kind of process leads to a higher pressure on the transport prices and thereby boosts wage dumping.

2.2 Innovations

Innovative logistic business models on the last mile are facing various partly conditional challenges and need to combine the aspects of the three dimensions of sustainability within one approach. On the ecological level those are environmentally and climate friendly transport and logistics processes, on the economical level profit and growth and on the social level fair payment, job stability and social security.

2.2.1 Electromobility

The integration of electromobility by the use of, for example, electric cargo bikes not just has positive effects on the carbon-dioxide footprint but at the same time helps improving the image of the companies using it. In combination with the use of micro depots as an intermediate between the depots of the CEP-companies and the cargo bikes, the disadvantages of the light electric vehicle (LEV), such as low loading capacity and driving range can be compensated. According to studies of the project 'Cycle-Logistics' it was demonstrated that cargo bikes have a huge potential to tackle these challenges as they can improve the image and general levels of cycling, replace over 50% of urban transport-related trips, as well as enhance air quality, safety levels, and life quality of urban areas. At present, however, this innovative solution is not fully deployed in any European cities' [10]. The studies of the project 'Ich entlaste Städte' shows that more and more companies tend to use cargo bikes to replace fuel-operated driving systems on the last mile. The initiative lends cargo bikes to companies for a euro a day for up to three months of testing. While and after the testing phase the companies are surveyed and after testing every fifth user bought an own cargo bike or considered buying one in the future [11].

2.2.2 Bundling

Furthermore, approaches such as bundling can lead to positive effects on sustainability dimensions of the last mile.

Bundling, in this context, refers to bundling of different products (newspapers, parcels, delivery services) as well as bundling of services for different clients (local retailers, GLS, publishers etc.). Furthermore, delivery and collection can be bundled in one tour, in which the term "pickup and delivery" is also used. Several aspects including disposition, tour planning, the staff and the delivery vehicles are to be considered in the process of bundling.

Ecological benefits due to the higher utilization of loading capacities and reduction of distances can result from bundling of transport in different ways. Moreover, on the economic level it can lead to time savings and cost reductions. As the better utilization of capacities takes advantages of economies of scale by dividing fix costs by more units, at the same time it furthers reduction in overall costs, environmental pollution and traffic congestion. With smaller shipment volumes, the costs per unit for bundling can be higher than for direct transport, since bundled transport leads to detours to different shipping and receiving points and causes additional downtime. In general, bundling has a positive effect on logistical sustainability and leads to improvements on economics, ecology and transport levels [12, 13].

For bundling, some product groups request special treatment, as for example some products are perishable or need special cooling, and the bundling results in longer transport times. Greater attention should be paid to the complexity of this instrument, and dynamic route planning must be geared to this. Uncertainties should be taken into account to route planning, so that planning steps must be continuously updated [14].

Dynamic data are, e.g.:

- **Customers**: Not all customers are available at the time of planning, as some of the customers or orders only become known as the tours are executed.
- **Service Time**: Customer events can extend the service time.
- **Delivery time window**: Customer can extend, shorten or shift the delivery time window.
- **Capacities**: Available capacities can vary, e.g., because of defective vehicles or sick employees.
- **Travel times**: Different travel times to consider at different times of the day, but also unexpected events such as accidents, weather conditions, etc.

To implement dynamic route planning innovative information technology is needed, while it has to be defined which data is required as input. Out of the input a certain output in shape of a planned route is generated with the help of an algorithm [15]. Inputs such as: Customer data, order data, driver data, vehicle data, stroke data, traffic information, distance matrix, product group etc. are needed for the data processing and the algorithm. The algorithm includes restrictions, input parameters, optimization objectives and secondary real-time data processing that is based on position data (geofencing), quantity (volume) update, tour control and changes in generation or distribution plans [15, 16].

2.2.3 Crowd Logistics

Crowd logistics leads back to the term crowdsourcing as a neologism of the terms "crowd" and "outsourcing", where "crowd" is defined as the mass of people or potential labor force, and "outsourcing" is the relocation of processes, functions, and duties to third parties [17].

Mehmann et al. define the term as follows: "Crowd-logistics refers to the outsourcing of logistics services to a large number of actors, whereby coordination is supported by a technical infrastructure. Crowd-logistics aims to achieve an economic benefit for all participants and shareholders" [17].

The technical infrastructure is usually used as a communication medium, such as a platform, which can be accessed in different ways (mobile phone, web browser etc.). The platform handles demand and supply for transport services, management processes and invoice processing.

On economical perspectives, benefits of crowd logistics are achieved from the Sharing Economy paradigm, a concept in which increased prosperity results from sharing goods or services between market participants. According to Mehmann et al. [17] crowd logistics enables new ways in logistics services and the improvement of existing logistics services in terms of volume, speed, and flexibility, which leads to win-win economic effects for all stakeholders. This can lead to significant growth of productivity on the last mile, for example. Recipients could be able to receive more convenient and flexible logistics services, while suppliers obtain economic benefits. The potential of crowd logistics can be observed in several successful companies, such as the Taxi Service Uber and its freight-on-demand platform Uber Freights. Certainly, it is increasingly seen that existing labor market models are leading to disadvantages of employees, especially when it comes to the low-wage sector, as the crowd logistics [18].

Problems in this context are fake self-employment, falling wage levels, shifting responsibilities and resources to the employee, as well as softening of the protection against dismissal etc.

Looking at social potentials of crowd logistics, it could be used not to replace the existing workforce but to support it in high phases and has the great potential to strengthen the over-the-counter retail by integrating them into the crowd logistics approach and bring them back some of their market shares.

In summary, it can be stated that crowd logistics models have high economic potentials but are also fraught with risks. An essential focus for companies should be to develop competitiveness but at the same time being employee-friendly based on the crowd approach in terms of social sustainability [16, 19].

2.2.4 Sustainable Customer Relationship Management (SusCRM)

Sustainability in CRM is discussed from several different aspects to date. As this approach features the notion of CRM, which focuses on customer relationship management or relationship marketing strategies, and the term ‘sustainable’. SusCRM can be understood as any CRM methodology which supports a continuous development of inter-business and customer relation while matching ecological, social and economic values for involved parties and as well relevant third parties. However, sustainability is growing more and more in importance in general within CRM research due to the immediate and direct impact on humanity from several perspectives that influences most meaningful things to mankind like, for example, life or a planet to live that has no substitution. Sustainability tends to be the main topic as it emphasizes since there would be missing essential parts of social and associated economic life without a viable environment. According to this, sustainability is an instrument of CRM supporting the process of identification, differentiation, interaction, and customization for customer clusters of current business cases. One of the key achievements of a successful CRM strategy is the creation of loyal customer groups for a sustainable business model. SusCRM strategies can lead to determine sustainability values and targets as key differentiators among customers and on the other hand present a designed CRM process which increases loyalty to the core value in a specific reward-based concept, which motivates its audiences to act more sustainable through the business model [16, 20].

3 Case Study Within the Project NaCl

3.1 Chances of NaCl-Approach

Within the "NaCl—Sustainable Crowd Logistics" project, an innovative and sustainable logistics system for last mile logistics was developed based on a crowd logistics approach. The project is currently planned in an inner-city test field in Bremerhaven shortly before being piloted.

Significant features of the designed logistics system are on the one hand the positive ecological effects using locally emission-free and traffic-sparing, electric cargo bikes. On the other hand, it has interesting economic potentials for the logistics service provider, as it is more elastic compared to conventional systems of CEP service providers based on combustion vehicles, especially in the personnel sector due to the lack of driving license and low qualification requirements.

The employment of student crowdworkers enables the delivery to be handled with maximum flexibility, as well as improving the working conditions of permanent parcel deliverers by taking over excessive parcel deliveries.

Another special feature of this project is that within the framework of a "combined delivery" the transports are bundled about both the delivered products and the logistics service providers commissioning them. As a result, logistics will be significantly more efficient and thus, more social and environmentally friendly. However, this makes the requirements on the system, both on an organizational and a technical level, more complex.

As part of the project approaches for Sustainability Customer Relationship Manage-ment (SusCRM) [20] were developed. These strengthen sustainability as a central marketing factor and exhibit gamified incentive systems for participation in a sustainable logistics system at the levels of logistics customer (B2B) and end customer (B2C) as well as at the employee level.

3.2 Conceptional Components

Partners of the project coordinated by the Bremerhaven University of Applied Sciences are the companies Rytle GmbH and the Weser Eilboten GmbH, an associated company of the Ditzen business group in Bremerhaven.

The logistics concept is based on the last mile logistics system of Rytle. This is a highly efficient system consisting of mobile hubs working as micro-depots, electric freight bicycles (MovR and Triliner) with an exchangeable transport box for the products to be delivered, and a digital infrastructure.

The digital infrastructure offers an innovative platform for networking between the software and hardware components involved in the logistics concept. Protected data on the products, customers and tour planning from the software of the logistics service provider Weser Eilboten are forwarded to the software of Rytle via this infrastructure. An optimized route planning for the last mile then is created for the crowdworker.

The main component is the driver app provided by Rytle, which was adapted for use by the crowdworkers as part of the NaCl project. The crowdworker is assigned to a cargo bike and a transport box via the app. It supports the crowdworkers in handling of transport and instructs them in the operating sequence of the customers. The driver app also supports pickup and delivery and the specification of customer time windows.

In addition to the driver app, the Rytle software also provides an app for business customers and end customers, in which shipments can be commissioned, tracked and evaluated. Additionally, to locking options for MovR, BOX and HUB, the app offers over 200 other use cases [21].

4 NaCl Model

For the implementation and testing of the NaCl model, a regionally based logistics service provider is required, who can take over capacities of other logistics service providers in addition to their own customer base. In order to be able to manage these additional tasks, crowdworkers will support the already existing staffing. A special feature of the pilot scenario is that the crowdworkers are mainly recruited from students of the Bremerhaven University of Applied Sciences. This focusing addresses a group that sets priority on a flexible working relationship, but at the same time is receptive to incentives through participation in an ecologically sensible system. It must be ensured that the permanent employees are working at 100% capacity and that the crowdworkers are only called in if there is an overload. Employees on vacation must be replaced, the Christmas business requires significantly more employees and cases of illness can be compensated. It is also possible to reduce overtime hours of the permanent staff quickly and easily. Consequently, this means a stable planning base for permanent employees with unlimited contracts and less stress.

The information of individual data streams, for example of tour planning, customer requirements, the specifications of the goods and the delivery route, must be managed transparently in one system (Fig. 1).

This project fixes that the crowdworkers will deliver entire boxes and moreover pickup and deliver parcels of the business customers of the regional stationary retail during the pilot phase.

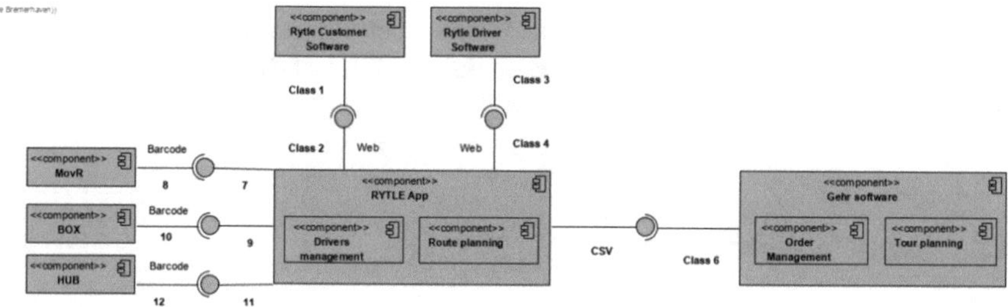

Fig. 1 Component diagram interfaces

In the area of SusCRM, the purely economic dimension of CRM is expanded to include an ecological and a social dimension in the sense of sustainability. Therefore, a SusCRM approach of the mobility domain is transferred and adapted. The main goal is the motivation especially of customers for a more sustainable consumption [20].

Various approaches will be pursued, for example, a consignment-related emission calculation is planned to compare air pollution on the last mile, which is on the one hand, based on the conventional delivery method using diesel internal combustion engines, and on the other hand, the load caused by the use of load bicycles. Also, concrete incentive systems are to be developed so that customers decide in favor of sustainable delivery, and students are encouraged to become part of this sustainable project. The focus here will be on social media and advertising campaigns as well as the university's communication channels. The sustainability goals of the project are analyzed by means of an evaluation and acceptance among the population, permanent employees and crowdworkers is recorded.

4.1 *Crowd Logistics and Work Situation*

The NaCl model provides a staffing structure that consists in part of crowdworkers. The crowdworkers will only be mobilized in peak times when permanent staff capacities are exceeded. The goal is not to replace the current staff, the crowdworker should relieve the staff and thus generate a significant overall improvement of the work situation with less overtime and reduced stress situations. This leads to a higher level of motivation among the permanent employees and to less fluctuation. The usage of cargo bikes has a distinct advantage that driving does not require a driver's license. The main drive of the crowdworker should not only be earning money but also to contribute to a cleaner environment. Delivering should be a flexible and lucrative part-time job for most drivers. Due to the bundling of different cargo and products and to the pickup and delivery of goods, the job confers more responsibility. The bundling will also reduce the costs of the logistics service providers. The cost reduction and higher responsibility will lead to higher wages of the regular driving staff [16].

4.2 Logistical Processes

In this part, we will take a close look at the logistical processes of the NaCl model. The logistical processes describe the basic logistical process, the selection process to find a crowdworker, the delivery process from the driver's perspective and the pickup and delivery process.

The general logistical process is shown in the following activity diagram (Fig. 2).

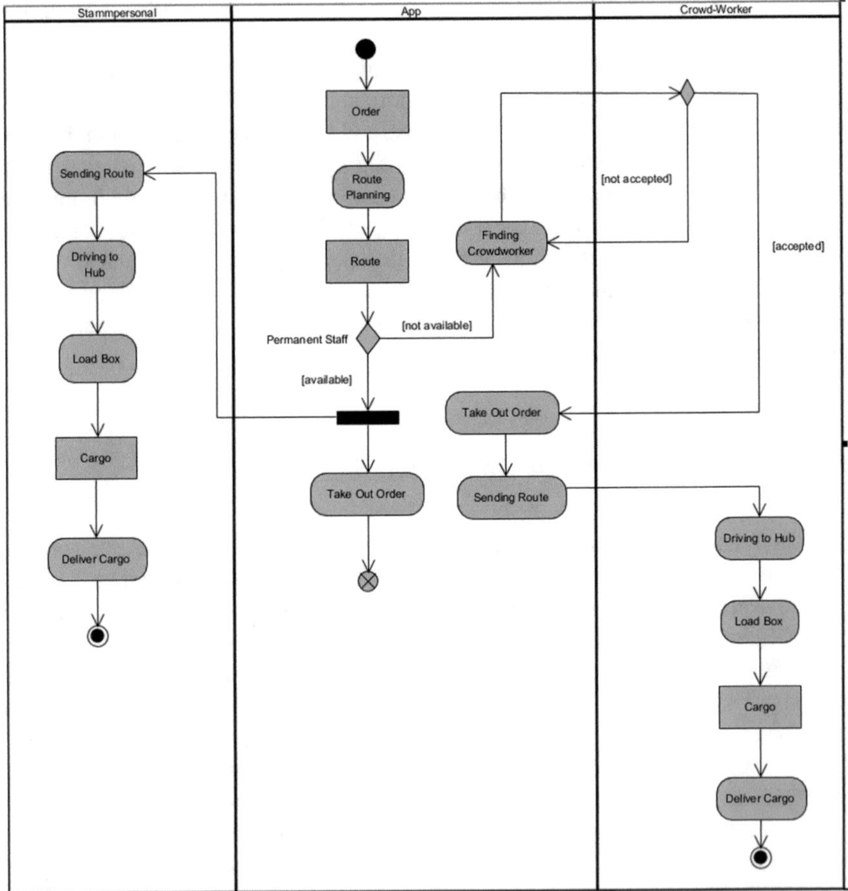

Fig. 2 General logistical process

Weser Eilboten GmbH creates its orders via an application and a software tool and merges them into a tour. If sufficient permanent staff is available, the tour will be delivered by them. Otherwise a crowdworker will be contacted to fulfill the orders. In this case one of the crowdworkers accepts the tour, which then is removed from the order pool. The delivery is made by cargo bike. The crowdworker picks up the cargo from a hub and delivers the goods [16].

In detail the crowdworker selection process "Finding crowdworker" is done through the Rytle app and is based on 3 criteria (Fig. 3).

1. Status:
   This means that only "standing by" crowdworkers receive a push notification. Standing by means that the crowdworker is logged into the driver app at the beginning of its standby time and switched to "active".
2. Distance:

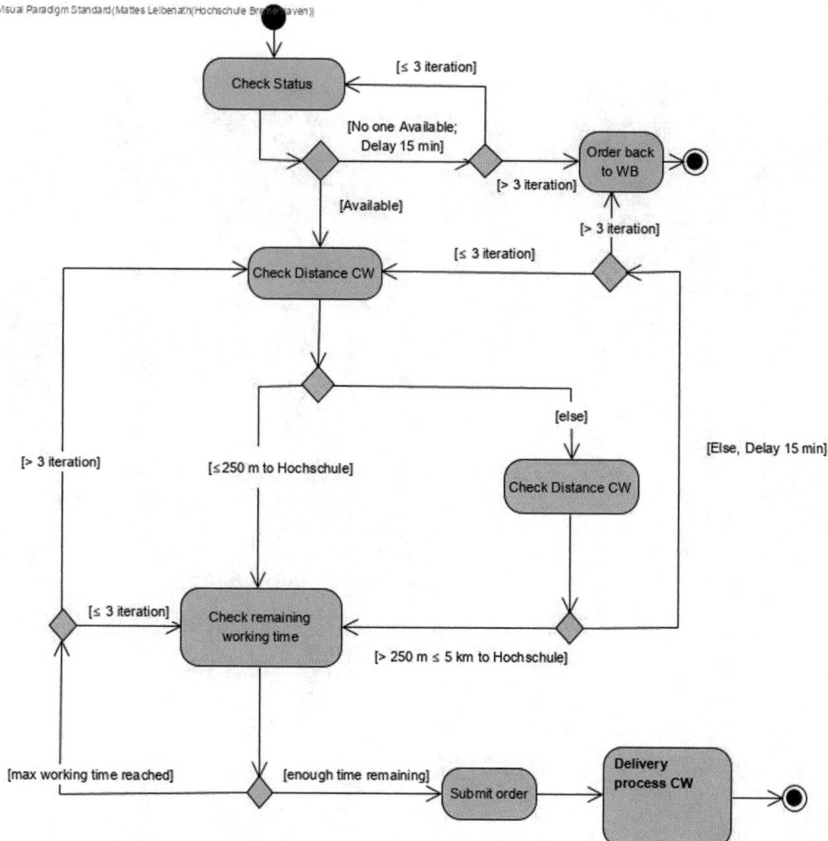

Fig. 3 Selection process crowdworker

Crowdworkers who are located less than 250 m from the university should be given preference when placing orders. If no crowdworker is available in this radius, it will be expanded to up to five kilometers.

3. Available working hours:
   Since student crowdworkers have a limited monthly working time, it is important to check whether the crowdworker assigned to the order does not exceed his workload.

If the crowdworker checked does not meet one of these criterias, another crowdworker will be checked after a delay of 15 min. This process is carried out in three iterations. If no crowdworker is found during this period, the delivery request returns to Weser Eilboten.

The delivery process from the driver's perspective is shown in the next figure (Fig. 4). After accepting an order, the crowdworker goes to the cargo bike (MovR or Triliner), which is available on the university campus. The student then scans the MovR using the driver app, which assigns it to a MovR ID. The lock is opened via

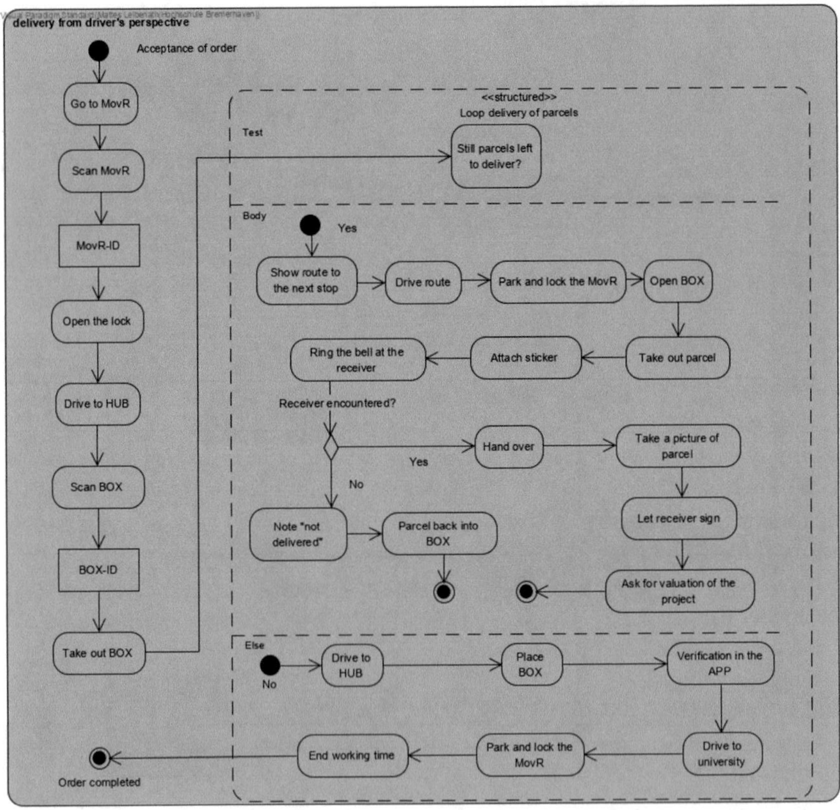

Fig. 4 Delivery from the driver's perspective

Bluetooth so that the crowdworker can now drive to the HUB to load the parcel carrying box. There is also an optional cargo bike (Triliner) at the university for pickup and delivery orders from the regional retail. The correct BOX is scanned at the HUB, a BOX ID is assigned, and the BOX is attached to the MovR. The delivery begins, while the routing is done via the mobile application.

After the first delivery is done, the crowdworker takes a photo of the package and saves it in the driver app. Now the customer has to sign it. If the customer is not present, a notice saying "not picked up" must be stored in the app and the package is placed back in the BOX.

This process repeats until all goods have been delivered. As a last step, the crowdworker returns to the HUB, leaves the BOX and confirms this process in the driver app. The MovR must be driven back to the university, where it is parked and locked. At the end the crowdworker confirms the whole order and completes his working hours.

Another important process within the NaCl project is pickup and delivery for business customers (see Fig. 5). A B2B customer can request a pickup in the customer app. The RYTLE software then tries to assign this pickup to a crowdworker who is already on a delivery tour. He receives a push notification and can accept the pickup. Afterwards the pickup will be inserted into the route as a tour stop. If no driver is on the road or the crowdworkers on a tour do not accept the pickup, it will be sent as a new order to all active crowdworkers via push notification. If no crowdworker accepts the order (e.g. due to the time or a lecture), there is a possibility that a driver of Weser Eilboten takes over the delivery of the pickup. The regular driver also delivers the pickup by cargo bike.

4.3 SusCRM Approaches and Incentive Systems

Within the framework of the project, incentive systems for participation in the sustainable logistics concept have been developed. At the employee level several concepts of gamification have been modeled and incentive ideas for the acquisition of drivers. Within this topic the biggest incentive should be that the crowdworkers support an environmental project with their commitment. In addition, the salary is an important factor within the acquisition process. Important incentives for the student crowdworkers are the high flexibility of the job and the participation in an innovative and modern project. Finally, the lack of a driving license increases the elasticity of the crowd approach. Communication of the benefits and special features of the sustainability logistics system motivates the crowdworkers to be a part of the active "Rider-Community".

Gamification is a concept in which game elements are integrated into non-game activities and applications. The goals of this transition are to influence user behavior and motivation. Also, loyalty is to be increased, guaranteeing the sustainability of the app [22]. Gamification can be used effectively, especially with apps whose basic tasks are monotonous and repetitive, and significantly improve commitment and

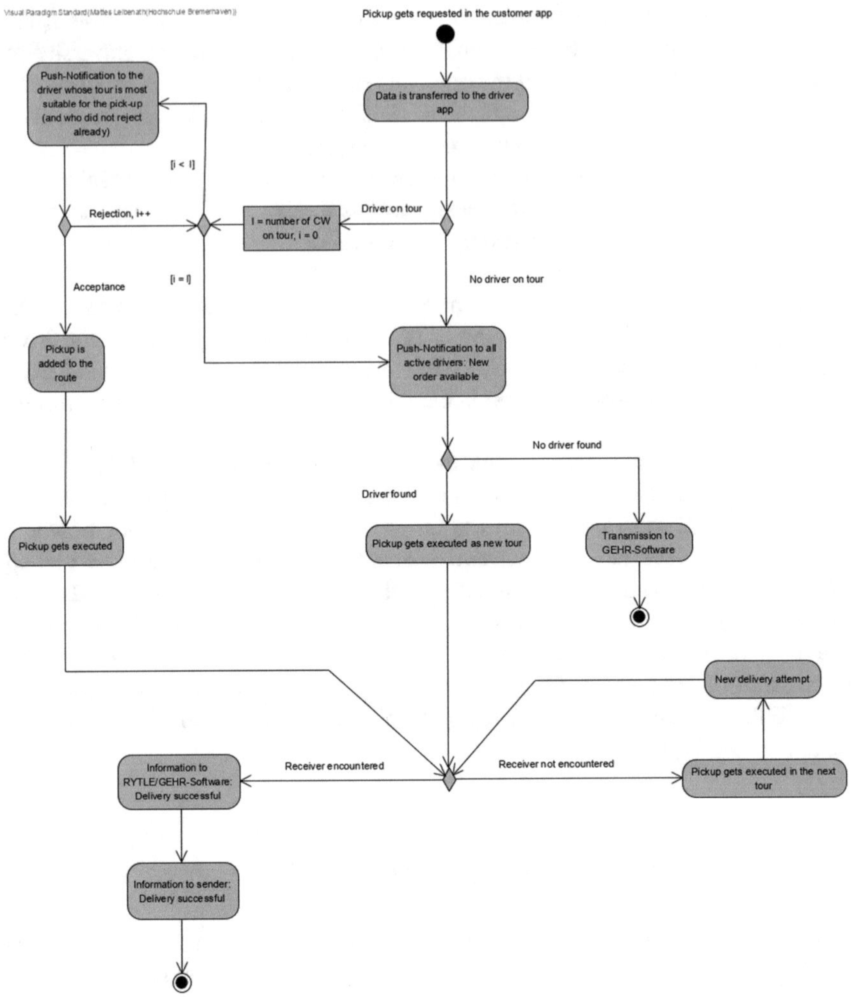

Fig. 5 Pickup and delivery

motivation. The process of implementing gamification elements can be subsumed into the points of challenge, reward, theme, and progress [16, 23].

For successful gamification within the framework of the NaCl project, several gamification ideas have been developed for the implementation in an app to encourage drivers to drive for a long time. At the same time, the incentive systems should increase the quality of the transport service. Within a level system of achievements, tasks and rewards experienced people shall be retained. Achievements are badges that are awarded to the driver after reaching specific milestones. Table 1 shows an example for this kind of gamification.

Table 1 Gamification system

Achievement	Task	Reward
Town expert	Have delivered a package in each district of Bremerhaven!	Golden flag for bicycle
Sportler	Drive 200 km by bike!	Race bar
Profi	Reach level 12!	Golden App layout
All for one!	Promote a friend for the App!	Extra big bicycle bell

These competition-based incentives with city and district-specific rankings would be a useful indicator for drivers of how they perform compared to their colleagues.

At the customer level, an approach for a SusCRM was developed. To achieve a high level of efficiency of the SusCRM approaches, it is necessary to set up an own CRM system, which allows customer relationships to be managed and strategies to be quickly adapted or created. Different incentive systems must be implemented so that the customer pays attention to sustainability in his consumer behavior. In the SusCRM approach, the distinction is made between information-based, social, and reward-based incentives, as well as gamification-based incentives are possible [20].

Within the NaCl project, these approaches were transferred to the last-mile-delivery domain. Specific use cases are shown in the following diagram (Fig. 6).

Information-based incentives are all those incentives which lead to a change in the customer's purchasing behavior by providing information to the customer. The pure providing of information is not enough for a change of behavior, so it must be complemented by attractive options and further incentives [20]. An app is intended to provide the customer (also the dispatcher) with information on the sustainability of the transport options.

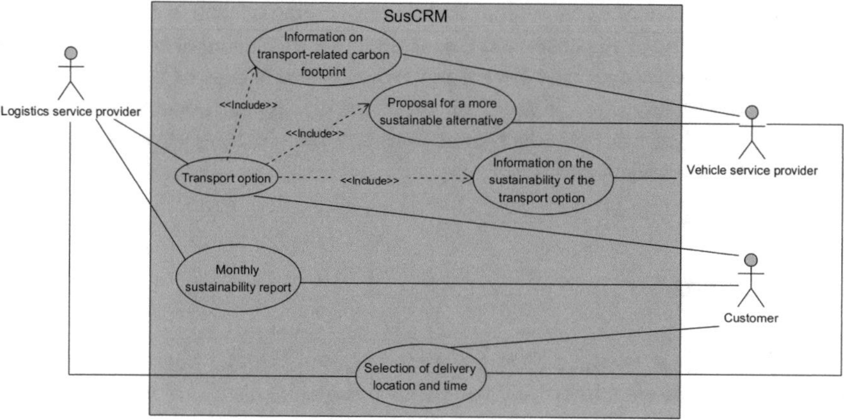

Fig. 6 SusCRM approaches on the customer side

Table 2 Incentive systems and marketing of sustainable logistics

Incentive system type	Customer loyality
Information-based incentives	• Re-use of packaging • Hint to the number of articles in the shopping basket • Hint to sustainable delivery option • Proposals for more sustainable products (less carbon footprint, label "Fair Trade")
Gamification-based incentives	• Carbon footprint comparison between the existing and a virtual shopping basket
Social incentives	• Sustainability report of the own buyer behavior • Share in social networks
Reward-based incentives	• Credit for no returns • Bonus for new customer acquisition • Selection delivery time and delivery site • Successful delivery in the first delivery attempt • Rebate system for new registration

With social incentives, the motivation for changing behavior arises from the expected reciprocity. This describes the human effort to increase his value in the community to satisfy the need for social attention [20].

Reward-based incentives involve material and immaterial incentives to influence the purchasing behavior of end customers. Material rewards can be divided into monetary rewards, such as credits and discounts, and non-monetary rewards, such as vouchers. Immaterial rewards do not influence behavior through material value but relate more to social incentives [20]. Table 2 shows an overview about possible incentive systems for customers.

Furthermore, it is also possible to agree on virtual delivery points with the delivery and the customer's location based on the current location of the cargo bike. This requires a high level of dynamic tour and route planning as well as communication with the customer. For the customer, the advantage is the saving of additional routes because he has to pick up his delivery at the post office in case of the unsuccessful delivery. The delivery also is faster. This offers the dispatcher economic and ecological advantages such as the reduction of journeys, the bundling of parcels, and the reduction of CO_2 emissions [16].

4.4 Pilot Phase

The project is about to start a three-month test phase. The pilot phase is used to test the concept of sustainable last-mile logistics using students as crowdworkers. The sustainable logistics solution will be tested in a defined delivery area in the center of Bremerhaven. Depending on the location of the business customers, the test field can be expanded flexibly during the pilot phase.

On the technical side, one MovR and two Triliner of Rytle will be used during the pilot phase. The driver app adapted to the project will be tested by the crowdworkers as well as the customer app that is only available to business customers in the event of pickup and delivery. The two applications may have to be adjusted during the test run if the new operating conditions require so. The pilot phase is documented via an evaluation study and the tour data will be monitored using the cargo bike telematics system.

Another important point during the pilot phase is the achievement of a high level of attention for the project. It must be examined how high the acceptance of the sustainable delivery method is according to the inhabitants of Bremerhaven. In order to be able to assess this, the customer is asked for a brief assessment of the project and the sustainable delivery as the goods are handed over. The NaCl website also provides evaluation options for end customers and business customers. A diary integrated into the project website about the progress of the pilot phase can create opportunities for participation in sustainable last mile logistics.

5 Conclusion and Future Outlook

In view of the diverse effects that on-demand platforms and increasing e-commerce have on last-mile logistics, the NaCl approach presented here can be a correct step towards an ecologically, socially and economically compatible logistics system.

In addition to reducing emissions and traffic jams in the city centers, relieving regular delivery drivers through well-paid student crowdworkers and bundling goods deliveries, the logistics system also has a positive impact on stationary retail businesses.

Particularly in view of crisis situations, such as those currently shaking the world (SARS-COVID-19), the sustainable logistics system offers stationary retail an alternative to realizing deliveries and disadvantaged people a chance to be supplied with important goods.

Acknowledgements The NaCl project is funded by EFRE within the program “Applied Environmental Research (AUF)” of the city of Bremen.

References

1. Gevaers, R., Van de Voorde, E., Vanelslander, T.: City Distribution and Urban Freight Transport: Multiple Perpectives, Characteristics and Typology of Last-Mile Logistics from an Innovation Perspective in an Urban Context. Edward Elgar Publishing, Cheltenham (2011)
2. Pwc: https://www.pwc.de/de/transport-und-logistik/pwc-studie-aufbruch-auf-der-letzten-meile.pdf. Last accessed 29 April 2020
3. Pwc: https://www.pwc.de/de/handel-und-konsumguter/studie-total-retail-2017.pdf. Last accessed 29 April 2020

4. OC&C Strategy Consultants: Endspurt – Der Wettkampf auf der letzten Meile. (2017)
5. (Bundesverband Paket und Expresslogistik e. V. (BIEK): KEP-Studie 2016 – Analyse des Marktes in Deutschland (2016)
6. BMU: https://www.bmu.de/fileadmin/Daten_BMU/Pools/Broschueren/klimaschutz_in_zahlen_2018_bf.pdf. Last accessed 29 April 2020
7. BMU: https://www.bmu.de/fileadmin/Daten_BMU/Download_PDF/Klimaschutz/klimaschutz_in_zahlen_klimaziele_bf.pdf
8. https://www.forbes.com/sites/solitairetownsend/2018/11/21/consumers-want-you-to-help-them-make-a-difference/#2eecd2066954. Last accessed 29 April 2020
9. Pwc: https://www.pwc.de/de/transport-und-logistik/pwc-studie-aufbruch-auf-der-letzten-meile.pdf. Last accessed 28 April 2020
10. https://cyclelogistics.eu/about. Last accessed 7 July 2020
11. DLR: https://www.dlr.de/content/de/artikel/news/2018/4/20181126_projekt-ich-entlaste-staedte-zwischenbilanz.html. Last accessed 28 April 2020
12. Fleischmann, B., Kopfer, H.: Systeme der Transportlogistik. In: Tempelmeier, H. (ed.) Begriff der Logistik, logistische Systeme und Prozesse, pp. 17–28. Springer Vieweg, Berlin (2018)
13. Wiese, J.: Slow Logistics – Eine simulationsgestützte Analyse der ökonomischen und ökologischen Potentiale der Sendungsbündelung. In: Sucky, E. (eds.) Logistik und Supply Chain Management, vol. 15. University of Bamberg Press, Bamberg (2017)
14. Richter, A.: Dynamische Tourenplanung. Modifikation von klassischen Heuristiken für das Dynamische Rundreiseproblem (DTSP) und das Dynamische Tourenplanungsproblem (DVRP) mit der Möglichkeit der Änderung des aktuellen Fahrzeugzuges. Dissertation. Technische Universität Dresden, Dresden (2005)
15. Anschütz, S.: Dynamische Tourenplanung im Teilladungsverkehr. Schlussbericht. Initions Innovative IT Solutions AG, Hamburg (2011)
16. Wagner vom Berg, B. et al.: A sustainable CRM approach to a crowd sourced last-mile logistics platform (NaCl). EnviroInfo 2019, Kassel (2019)
17. Mehmann, J., Frehe, V., Teuteberg, F.: Crowd-logistics—a literature review and a maturity model. In: Kersten, W., Blecker, T., Ringle, C.M. (eds.) Innovations and Strategies for Logistics and Supply Chains, pp. 117–145. Epubli GmbH, Hamburg (2015)
18. Abel, J. R., Florida, R., Gabe, T.M.: Can Low-wage workers find better jobs? In: FRB of New York Staff Report No. 846. https://ssrn.com/abstract=3164963. Last accessed 14 May 2019
19. Wagner vom Berg, B., Moradi, M.: Sustainable labor conditions by Gig-economy—case study: sustainable crowdlogistics (NaCl). In: Weizenbaum Conference, Bremerhaven (2019)
20. Wagner vom Berg, B.: Konzeption eines Sustainability Customer Relationship Management (SusCRM) für Anbieter nachhaltiger Mobilität. Shaker Verlag, Aachen (2015)
21. Rytle GmbH Website, https://rytle.de/. Last accessed 3 July 2019
22. Law, F.L., Kasirun, Z.M., Gan, C.K.: Gamification towards sustainable mobile application. In: Harun, M.F. (ed.) 5th Malaysian Conference in Software Engineering (MySEC), pp. 349–353. Piscataway, IEEE (2011)
23. Flatla, D. R., Gutwin, C., Nacke, L. E., Bateman, S., Mandryk, R. L.: Calibration games: making calibration tasks enjoyable by adding motivating game elements. In: Pierce, J., Agrawala, M., Klemmer, S. (eds.) Proceedings of the 24th Annual ACM Symposium on User Interface Software and Technology (UIST'11), pp. 403–412. ACM Press, Santa Barbara (2011)

Environmental Modelling, Monitoring and Information Systems

Algorithmic Treatment of Topological Changes Within a Simulation Runtime System

Jochen Wittmann

Abstract Many applications in the field of environmental simulation are not limited to the dynamics of one-dimensional inventory variables, but additionally try to describe the spatial dimension of the investigated objects with their dynamic changes. The paper is based on a general specification level for such topological changes of geo-objects using graph replacement systems and shows how such a specification can be integrated transparently and consistently as an extension of the basic algorithm for combined, discrete–continuous models. The topological events are analysed in their semantics and their treatment is presented as an additional, modular algorithm part. Problems caused by this kind of specification and processing are uniqueness problems when several events occur simultaneously as well as the stability of the model specification in case of event cascades. Both problem areas are discussed with regard to their importance for the semantics of the model specification and the interpretation of the simulation results.

Keywords Modeling · Simulation · GIS · Spatio-temporal · Model specification paradigm · Simulation runtime system

1 Motivation: Simulation and Geoinformation Systems

The requirements for analysis, modelling and simulation of dynamic processes have changed fundamentally in recent years. On the one hand, there has been an increasing spread of smartphones with automatic position determination via GPS satellites on the data acquisition side, which has become the standard even for the consumer sector, and on the other hand with free, convenient and fast access to geographical maps and images, e.g. via the web GIS Google Maps [1] or OpenStreetMap [2]. The approaches to modelling can now be geographically differentiated with high spatial resolution (examples e.g. in [3]). On the part of computer science, this trend (foreseen by Goodchild already in the late 1990th [4]) is supported by the concepts of

J. Wittmann (✉)
Hochschule für Technik und Wirtschaft Berlin, University of Applied Sciences, Wilhelminenhofstraße 75A, 12459 Berlin, Germany
e-mail: Jochen.Wittmann@HTW-Berlin.de

A. Kamilaris et al. (eds.), *Advances and New Trends in Environmental Informatics*, Progress in IS, https://doi.org/10.1007/978-3-030-61969-5_8

object-oriented programming languages and individual-based modelling techniques, which make it possible to handle a large number of objects or individuals, even spatially differentiated ones, in a relatively simple and descriptive way (see e.g. [5]) or an introduction to object-oriented programming [6]. While the c space-time data is essentially solved by corresponding object-oriented database concepts, the specification of dynamic models with reference to space and time is difficult, as comprehensively described in [7]. As a conclusion, a specification level is required there, which allows even complex dynamic processes to be specified on a meta level that is easily accessible even for the programming language layman and without in-depth programming knowledge.

This requirement is further investigated in [8] and leads to the idea to transfer the approach of graph grammars to the specification problem in geoinformatics. A concise and little mathematical summary of these considerations is given in the second section of this manuscript. However, the two preliminary works only deal with the meta level for the specification of topo-logical changes of geo-objects, but not with the implementation of this specification on the level of the implementation of the simulation runtime system. This step from the specification to an implementation concept is covered in detail in this paper.

2 Topological Changes Specified by Graph-Grammars

The following paragraph is a shortened and simplified version of the more detailed article [8] to introduce the basic idea of the dynamics specification for topological objects, for which the implementation concept will be developed in section three.

2.1 Continuous, Discrete and Combined Model Specification

Within the framework of simulation technology, the specification of the dynamics of system components is formalized either in the form of so-called continuous models or discrete models [9]. In addition, the third class is the set of combined models, which, however, do not contribute any new aspects methodically and only describe the combination of discrete and continuous approaches in one model.

Continuous models describe the system behavior by specifying difference or differential equations for the system variables. This allows to describe changes in levels or movements of objects in space and time.

On the other hand, only the so-called event is available as a construct for the specification of changes in discrete models: An event consists of a condition of occurrence, which specifies at which point in time the event is active, and the description of the changed system state (by the event). The status change described by the event takes place at the time of the event without any time consumption, and therefore takes effect immediately and suddenly. Put simply, events describe before/after rules that are executed when the event condition is true.

2.2 *The Application to Geo Objects*

Applied to geo-objects the situation is as follows: In the case of a continuous dynamic specification (e.g. for a simple polygon as shown in Fig. 1 the problem arises that the individual determination points of the object can move with different dynamics (in the example velocity and direction) and thus, when the dynamic equations are processed, inconsistencies and shifts of the topological invariants can arise very quickly. In the example the original polygon becomes a topologically completely new object due to the intersection, which was probably not intended in this form by the specification of the dynamics.

If, on the other hand, the discrete modelling methodology is used, the shape of the polygon after the event would have to be explicitly specified as in Fig. 2 and it would be possible to see even before runtime that the new shape is a violation of topological rules.

This advantage of discrete modelling, which concerns the "after" state, so to speak, is now extended by the approach of a model specification using graph grammars to include the part of the event that corresponds to the trigger condition, the "before". The following paragraph will clarify this.

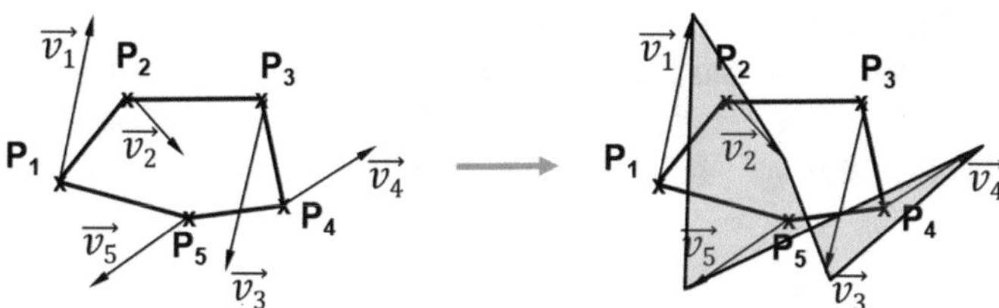

Fig. 1 Dynamics for a polygon by continuous model specification techniques

Fig. 2 Dynamics for a polygon by discrete model description techniques

2.3 Graph Grammars for Dynamics Specification

A mathematically complete definition of the so-called "graph grammars" will not be given here; based on the work of Chomsky [10], it is explained in detail by Schneider in [11]. Simplified, the central idea of the approach is the formulation of before-and-after rules for topological objects. However, these rules of a graph grammar do not describe the properties of the objects as a whole, but they are general rules for parts of a topology. If the pattern of the before-side of a rule is found at any point in a given topology, the rule can trigger and replace the before-structure found in the topology by the after-structure of the rule.

Transferred to the formalism of graph grammar, these replacement rules are the so-called "productions" of such a graph grammar.

Thus, the basic procedure how to work with the graph-grammar formalism can be described very simply with 4 steps:

1. create a set of productions that reflect the dynamics of the treated objects.
2. find in an existing graph a representation of the left side of a production and extract this left side as a subgraph.
3. in the subgraph given by the left side, execute the changes according to the right side of the production.
4. re-integrate the changed subgraph into its context of the existing graph.

Although these 4 rules clearly describe how to work with such a graph replacement system, they are firstly limited to exactly one replacement specified by a topological production and secondly leave completely open how the topological change is to be interpreted with respect to its behavior in the dimension time.

In order to solve these questions, the next section will focus on the integration of such topological graph replacement rules for topological objects in the implementation of a model specification for conventional (non-topological) combined models.

3 Treatment of Graph Grammar Productions in a Simulation Runtime System

3.1 Basic Algorithm for Simulation of Combined Models

Before the integration of graph substitution rules can be discussed, this section will explain the basic procedures for the algorithmic treatment of a combined, i.e. discrete–continuous, model specification. The explanations are mainly based on the methodology of systems theory and the work of Schmidt [12] and Eschenbacher [13, 14]. Alternative formalizations can be found in [15] and [16] by Zeigler and Uhrmacher.

The starting point for all considerations is a classification of the system quantities occurring in three classes:

(a) Independent input variables:
These are the information flows at the system boundary. The value of the input into the system of these input variables must be known at all times. It can be constant or determined by a calculation rule, but must not be dependent on the values of the other two types of system variables (dependent variables or state variables).
Example: (a) mean sunshine duration as constant or (b) sunshine duration depending on the season calculated by an adapted sine function with the day of the year as the only input.
(b) Dependent quantities:
The value of the dependent variables is determined at any point in time solely by the current values of the input variables and the state variables. For a better readability of the equations, the current values of the other dependent variables of the model can be additionally allowed for determination.
Example: State variable is the stored energy of a physical body, from which the temperature is derived as a dependent variable.
(c) State variables
State variables represent the memory variables of the system, which is why an initial value is also necessary for every state variable. Only for state variables dynamic relations can be specified, which are either formulated as continuous time progress in the form of differential equations or for discrete time progress in the form of events (with event occurrence condition and specification of the subsequent state for the respective model variable).

On the basis of this classification the basic algorithm shown in Fig. 3 can be established (programmers may excuse the use of goto, it serves here for better readability!) A detailed explanation cannot be given in this context, in particular, reference should be made to the work of Eschenbacher already cited. Rather, it should be asked and discussed at this point how the topological changes specified by the productions of a graph-replacement system can be integrated into this basic algorithm.

3.2 *Integration of Topological Changes into the Basic Algorithm*

With the argumentation carried out so far, the treatment of the topological changes seems to succeed obviously and consistently:

1. because of the before/after character of the graph grammar productions, their semantic treatment as events is obvious and consistent with a modellers' intuitive understanding.

Fig. 3 Basic algorithm for combined simulation

```
1. initialize
   t := Tstart;
   z(t) := z(Tstart);
2. set values of input variables
   x(t) := e(t);
3. update values of dependent variables
   y((t) := g(x(t), y(t), z(t));
4. execute state transitions
   if < event_condition_i >
     as long as event conditions are true ...
        then ...execute the events
             Execute(event_i);
             goto 3;
        else  check end criterion
             if ( t== Tstop)
               then        STOP
               else        execute time-step
                     z(t+ΔT) := f(x(t), y(t), z(t));
                                t := t + ΔT;
             goto 2;
```

2. topological objects as a whole are to be understood as state variables that change their state exactly when the current state is in agreement with a before/after topology of a production from the grammar. The after-state of the object is determined by applying the production. The change takes place without any time delay.

Consequently, the topological events could be processed on a purely formal basis and without any special treatment by the described basic algorithm. However, it should be kept in mind that topological changes affect more complex objects and usually also change more complex structures. Exactly for this reason, a "set of changes to an object" has been combined in one production to obtain consistency and topological properties in complex dynamics. From this consideration two conclusions should be drawn:

1. First, the topological events do have a different quality than the conventional ones and should therefore be treated separately.

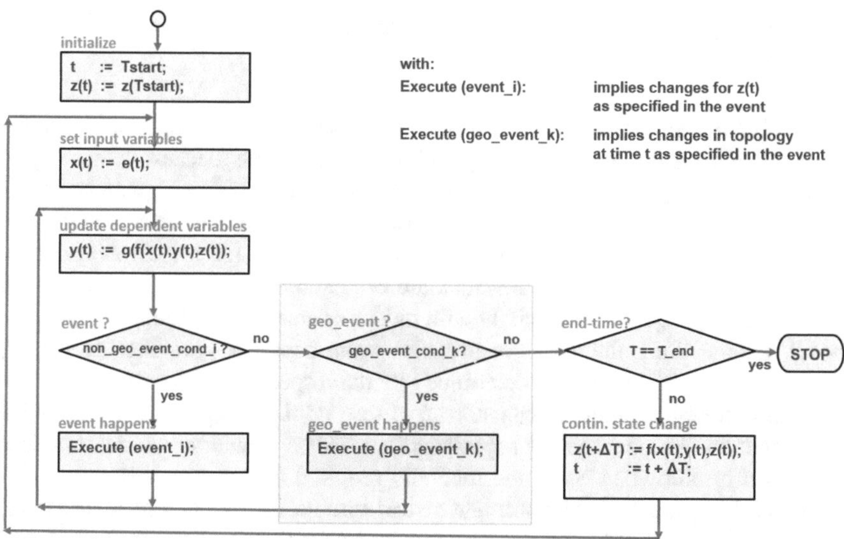

Fig. 4 Basic algorithm with topological events

2. Secondly, the possible triggering of the topological events should only be checked when the system state is stabilized, i.e. when there are no further state changes due to conventional events.

If the two decisions made in this section are combined and the resulting algorithm is presented as a flowchart, the picture shown in Fig. 4 emerges, which completely and consistently represents the required integration of the topological changes.

It should be noted that the model state can change after the execution of geo_events in such a way that, firstly, the dependent variables must be updated and, of course, new triggering conditions of further events may have become true.

3.3 Problems Caused by This Approach

Of course, the described procedure not only has the described advantages, but also has disadvantages and brings new problems. Two of the most important problem areas should be mentioned here because they are of particular importance for the semantic interpretation on specification level.

3.3.1 Problem of Sequence Dependency When Triggering Several Events Simultaneously

This problem is already known from conventional events, but should not be ignored when transferring the event handling to topological events because of its importance for a correct model specification. As can be seen in Fig. 4 with the algorithm, events can of course only be processed individually and sequentially. This implies a sequence in which event conditions are checked and if the conditions are met, the corresponding state transitions are executed. Since events do not have a naturally given sequence, especially with object-oriented and modular-hierarchical model specifications, the sequence in which event conditions are checked usually depends on internal conditions determined by the respective implementation of the algorithm (compiler, internal data storage of the simulation system, …) and is not transparent and influenceable for the specifiers. This can lead to non-deterministic simulation results when accessing mutually required resources at the same time. Such non-deterministic behavior might be undesirable, especially with regard to the reproducibility of simulation runs. An example: Due to two simultaneously active left sides of a production, a brownfield site can be converted into an industrial area on the one hand and into a residential area on the other hand.

There are two possible solutions: Either one accepts this indeterminacy and considers it as a stochastic influence on the simulation run, or alternatively one is stricter and interprets this conflict case as an error in the model specification, because it violates the rule of unambiguity and actually allows two alternative subsequent states to occur.

Regardless of the decision, it is important to note that implementation conditions permeate to the specification level, and that this must be taken into account when interpreting the results.

3.3.2 Problem of Event Cascades and Stability

A second problem also has to do with the possibility of simultaneous events. Again an example—admittedly very simplified—which exclusively uses the successive execution of topological events: An existing road network is extended at one point by a new connection. In the vicinity of the new road, this creates the prerequisite for the transformation into a new development area. Because of the attractiveness of the new development area, the plots are built on very quickly and the population is growing rapidly. However, the growing population makes the area unattractive, which in turn leads to the abandonment of the plots. All these separate steps could be mapped by topological events and in the end, as soon as the condition for the first event is fulfilled, a cascade of events would occur, which of course—that's just how the events are defined!—takes place at a single point in time without progress in the simulation time. In this case the processing in the simulation obviously does not meet the expectations and the desired semantics of the specification.

In these cases, it is important to ensure that the event conditions do not exclusively include topological attributes, but also other constraints (for example, the time of the last change of the topological object). The formalism of graph grammars does allow such constraints. It is therefore the task of the model specification and thus of the modeler to consider such event cascades already during model construction and to prevent them by appropriate constraints. Here again, automated support can be provided by automatically analyzing the given set of graph productions for possible chains and then allowing the modelers to recognize and solve the described conflicts before the simulation is executed.

4 Conclusion and Discussion

The present paper shows that the idea of using graph grammars for the specification of dynamic changes of topological objects does not only work on the specification level, where it represents an intuitively understandable and formally and algorithmically well treatable description approach. Rather, the formal specification given by the productions of a graph grammar can be easily and consistently transferred to the implementation level and integrated into the basic algorithm for simulation.

The main advantage of this approach is the fact that extensive consistency checks and plausibility checks can be performed on the specification level and solely on the basis of the given set of productions even before the model is run. In addition to the more defensive checks, as indicated in the previous subsection, more complex investigations of the rule set with the appropriate methods of AI are also conceivable here, an idea that can be extended to an automated model check based on the graph replacement systems specified.

As the discussion shows, the problems of simultaneity and stability known from the treatment of conventional events also occur in the treatment of topological events. Here, special attention is necessary but also possible for the model specification and again automatic support for finding event cascades. Special attention should be paid to the interpretation of simulation results concerning simultaneousness and stability, a problem however, already known from the treatment of conventional events that arises solely from the definition of an event as a change of state without time delay, and not from the use of the newly introduced graph grammar approach.

Experiences with smaller practical examples suggest that the approach is also suitable for complex models, whereby the strict formalization of the dynamics description in the form of graph grammar provides a very good basis for sophisticated methods of AI for an analysis and optimization of both the model specification phase and the phase of processing by the runtime system.

References

1. Google Maps, 5. 3. 2019. [Online]. Available: https://maps.google.com
2. OpenStreetMap, 5. 3. 2019. [Online]. Available: https://openstreetmap.org
3. O'Sullivan, D.P.G.: Spatial Simulation—Exploring Pattern and Process. Wiley, Chichester (2013)
4. Goodchild, M., Longley, P.: The future of GIS and spatial analysis. In: Geographical Information Systems: Principles, Techniques, Management and Applications, pp. 567–580. Wiley, New York (1999)
5. Ortmann, J.: Ein allgemeiner individuenorientierter Ansatz zur Modellierung von Populationsdynamiken in Ökosystemen unter Einbeziehung der Mikro- und Makroebene, Rostock: Dissertation am Fachbereich Informatik, Universität Rostock (1999)
6. Balzert, H.: Lehrbuch der Objektmodellierung. Spektrum, Analyse und Entwurf, Heidelberg (1999)
7. Zeigler, B.: Object-Oriented Simulation with Hierarchical, Modular Models. Academic, London (1990)
8. Uhrmacher, A.: Dynamic structures in modeling and simulation—a reflective approach. ACM Trans. Model. Simul. **11**(2), 206–232 (2001)
9. Schneider, H.: Graph Transformations—An Introduction to the Categorical Approach, Vorlesungsunterlagen (2019). [Online]. Available: https://www2.cs.fau.de/staff/schneider/gtbook/index.html
10. Chomsky, N.: On Certain formal properties of grammars. Inf. Control **2**, 137–167 (1959)
11. Wittmann, J.: Graph-Grammars to specify dynamic changes in topology for spatial-temporal processes. In: Schaldach, R., Simon, K.-H., Weismüller, J., Wohlgemuth, V. (eds.) Advances and New Trends in Environmental Informatics: ICT for Sustainable Solutions, pp. 131–148. Springer, Berlin (2019)
12. Wittmann, J.: GIS and simulation: survey and deficits in the specification of integrated models in 4D, pp. 49–58. In: AGIT 5-2019 Journal für Angewandte Geoinformatik, Berlin, Wichmann (2019)
13. Banks, J., Carson II, J., Nelson, B.L., Nicol, D. M.: Discrete—Event System Simulation. Pearson (2005)
14. Eschenbacher, P.: Entwurf und Implementierung einer formalen Sprache zur Beschreibung dynamischer Modelle, Erlangen: Dissertation Universität Erlangen-Nürnberg, Institut für Mathemathische Maschinen und Datenverarbeitung (1989)
15. Eschenbacher, P.: Konzeption einer deklarativen und zustandsorientierten Sprache zur formalen Beschreibung und Simulation von Warteschlangen- und Transportmodellen, San Diego, SCS Frontiers in Simulation (1993)
16. Schmidt, B.: Systemanalyse und Modellaufbau. Springer, Berlin (1985)

Citizens in the Loop for Air Quality Monitoring in Thessaloniki, Greece

Theodosios Kassandros, Andreas Gavros, Katerina Bakousi, and Kostas Karatzas

Abstract Air pollution may dictate the quality of the indoor as well as outdoor atmospheric environment. Due to its complicated nature, it is important for people to be able to identify how air quality (AQ) is interwoven to their everyday life, and how it is related to specific everyday activities. On this basis, and in the frame of the citizen science project URwatair (www.urwatair.gr) that was set-up in the Greater Thessaloniki Area (GTA), a number of AQ-related activities were designed. For this purpose, we employed a number of materials (questionnaires, low-cost AQ monitoring devices, on-line collaborative electronic workspaces) as well as methods (social media, hands-on workshops, gamified collaborative data processing), in order to engage citizens in the environmental monitoring and related knowledge extraction process. The aim of this study was to evaluate the methodology implemented in the URwatair project and to appraise results and conclusions extracted explicitly from citizens. The results indicate that citizens can identify main sources of air pollution in various situations and suggest everyday practices that can lead to lower exposure in high particulate matter concentrations.

Keywords Air quality · Citizen science · Internet of things

T. Kassandros (✉) · A. Gavros · K. Bakousi · K. Karatzas
Environmental Informatics Research Group, School of Mechanical Engineering, Aristotle University, Thessaloniki, Greece
e-mail: tkassand@physics.auth.gr

A. Gavros
e-mail: andreasga@gmx.com

K. Bakousi
e-mail: kbakousi3@gmail.com

K. Karatzas
e-mail: kkara@auth.gr

A. Kamilaris et al. (eds.), *Advances and New Trends in Environmental Informatics*, Progress in IS, https://doi.org/10.1007/978-3-030-61969-5_9

1 Introduction

Air pollution affects the health of millions of people, especially those living in densely populated urban areas. According to WHO, ambient air pollution accounts for an estimated 4.2 million deaths per year due to stroke, heart disease, lung cancer and chronic respiratory diseases; this mortality is due to exposure to small particulate matter of 2.5 microns or less in diameter (PM2.5) [1]. Road traffic, home heating, industrial emissions, shipping emissions, and other anthropogenic actions are the major sources of air pollutants.

As people spend a considerable amount of time indoors, indoor air quality plays a significant part in their general state of health. Whilst outdoor air quality is well monitored according to Ambient Air Quality Standards for the European Union [2], no such limits are specified for indoor environments. Moreover, people are increasingly spending more time commuting to and from work each day, and commuters' rush hour exposures are significantly influenced by mode of transport, route, and fuel type [3].

Citizen science is scientific research conducted by non-specialist citizens. The Internet and smart devices are making the collection, storage and manipulation of scientific data, resulting from the aforementioned research, more feasible. Citizen science initiatives with a focus on air quality commonly use low-cost measuring devices to learn more about local or regional air pollution and its sources. Such initiatives can produce useful information about local air quality [4].

Aspects of everyday behaviour of citizens while conducting their daily routine of activities may result in consecutive exposure to high air pollutants concentrations. Small alterations in such a daily routine could lower such exposure. On this basis, the active involvement of individuals, through citizen science projects, in identifying patterns of everyday life associated with high air pollution levels, can have an important impact on their knowledge and attitude and result in change of behaviour [5].

2 Materials and Methods

The URwatair project consortium consisted of two research teams and a non-governmental organization (NGO), while it had the official support of two municipalities of the GTA (municipality of Thessaloniki and municipality of Kalamaria). The project was divided into two activities which involved citizens in the monitoring of either air pollution or rainwater flooding incidents in the area, and in the co-development of best practices for related everyday utility. Concerning the quality of the atmospheric environment, the main goal was to support citizens in studying the quality of their breathed air, on the basis of scenarios reflecting their everyday life, and in the analysis and reasoning process based on collected data. The project was initiated at the beginning of 2019 and was planned to run until summer 2020.

The GTA is inhabited by 1,000,000 people and has a Mediterranean climate with hot, dry summers and mild, wet winters. Air pollution is among the most pronounced environmental problems of the area, where traffic, central heating, industrial emissions and natural emissions, together with urban morphology, result in high levels of pollutants like particulate matter (PM10-inhalable coarse particles as well as PM2.5 respirable fine particles), ozone, nitrogen oxides, carbon monoxide, etc. [6]. Due to the frequent exceedances of the European AQ guidelines concerning the concentration of particulates in Thessaloniki, and their abundance throughout the year in indoor as well as outdoor environments [7, 8], we selected this pollutant to be our goal in the frame of the URwatair citizen science project.

2.1 *Software and Hardware Related Materials*

At the initialization of the project, we used an online questionnaire in order to receive feedback from citizens concerning their interest and their knowledge on air pollution problems in the GTA. Moreover, we set-up an online collaboration environment for instant message exchange and data sharing based on the www.slack.com platform, in order to use it for continuous and dedicated communication and support towards the participating citizen scientists.

Concerning AQ monitoring, we employed low-cost sensors as they become more and more popular [9], especially for citizen science projects [10]. They are easy operated (even by citizens with low or no scientific background), recordings are uploaded on the Internet in order to create interactive maps using the data collected from the total number of sensors online, and some of them have built-in GPS to perform real-time geotagged AQ monitoring. Considering all these factors, in the frame of the URwatair project three different low-cost measuring devices were used to access the concentration of both coarse (PM10) and fine (PM2.5) particulate matter:

1. AirVisual Pro (https://www.iqair.com/): an air-quality device equipped with a quad-core chip and 3 GB RAM to perform control and processing. It uses the AVPM25b PM optical counter sensor that AirVisual co-developed, which is exposed to PM drew in the device by a small fan. A photo-sensor algorithm analyses the light refraction to output PM2.5 and PM10 concentration values. It uses a miniature, non-dispersive infrared (NDIR) sensor for CO_2 (S8, manufacturer: SenseAir AB), while it also includes temperature and relative humidity sensors. It employs a WiFi communication module and it makes data available via the AirVisual cloud (https://www.iqair.com/air-quality-map).
2. Sensebox home (https://sensebox.de/): an Arduino-controlled device equipped with the SDS011 optical counter sensor (manufacturer: Nova fitness Co., Ltd) for PM2.5 and PM10, an atmospheric pressure and temperature digital sensor (BMP280, manufacturer: Bosch Sensortec), a GPS,

and a WiFi and LoRa communication module, while data are available via the https://www.opensensemap.org/ .

3. Dylos DC1700 (www.dylosproducts.com): A device that is equipped with a proprietary particle counter, providing number of particles per volume mass between (a) 0.5 and 10 μm and (b) 2.5 and 10 μm. The device uses a small fan to draw in air and particles and funnels them through baffles molded into the case past the laser beam operating at 650 nm wavelength. The number of the particles corresponding to the two size bins are not estimated via a physical size detector, but via discrimination using an algorithm on the signal from scattered light. The device stores measurements locally and it can be connected to a PC via an RS232 port.

2.2 Methods

The current state of citizen science in Greece is not developed enough to have acquired a critical mass of citizens that would allow for a seamless initialisation of new citizen science initiatives. We therefore addressed the general public as well as citizen groups already active in environmental issues, making use of general as well as targeted invitations to citizen workshops through social and mass media.

In order to associate air pollution levels with citizens' everyday life, we defined four scenarios for indoor and outdoor AQ measurements and everyday utility recording via personalised daily reports diaries. The outdoor scenarios included the "Commuting" scenario as well as the "My Neighbourhood" scenario. The former focused on air pollution levels on daily routes, conducted by the citizens, using the same means of transport. The latter scenario aimed at recording the air pollution in neighbourhoods where the participants reside. Concerning the indoor AQ, the "Accommodation Space" scenario pertained to the particular room of one's house, where they choose to spend most of their day, like the living room. The "Cooking" scenario valued the various air pollution levels caused by different cooking techniques, and hence how each of them affects quality of life. To provide citizens with scenario-related guidance, we created event-driven diaries per scenario, targeting in correlating the measurements with everyday activities and practices that affect the quality of the breathed air.

Another methodological aspect addressed was the collection of feedback from participants, in accordance with the AQ measurements resulting from the scenarios that they applied, in order to allow them to process collected data and extract knowledge. For this reason, after the first two citizen workshops and citizen measurements rounds, a third citizen workshop was organised, involving a gamified problem-solving technique [11]: Citizens were supplied with filled event-driven diaries along with their corresponding data and timelines. Participants analysed the data and extrapolated about the main everyday activities that degrade AQ, while they brainstormed about possible solutions. Consequently, they were provided with blue and orange post-it notes and were asked to write down what they can detect as a problem (everyday activities that result in bad AQ) and hence proposals on possible solutions (actions to prevent AQ deterioration). Finally, citizens posted up their cards on a whiteboard where everyone could see, and the results of this procedure were openly discussed between them and the authors, arriving at common conclusions.

3 Results

The questionnaire on the perception and awareness level concerning air pollution in the GTA received responses from 21 participants. The majority expressed their interest in air pollution issues, their concern about the potential of improvement concerning AQ in the area, and their interest in being timely and in detailed informed about air pollution levels when it comes to everyday utility.

Coming to the online platform (www.slack.com), it was found to be convenient for communication and data sharing among participants, but as a non-mainstream tool for communication in Greece, visits declined through time and traditional communication tools were used (such as phone calls and e-mails) to technically support citizens in conducting their measurements. With 52 members on the platform, a total of 707 messages were exchanged, with 5.35 mean weekly active users and 53% of views in posts.

Most of the citizens committed to the agreed agenda, 48 citizens took part in one at least workshop, 33 of them followed one of the predefined AQ monitoring scenarios, while 22 of them delivered a diary and seven of them repeated the experiment delivering more than one diary. Finally, 21 citizens participated in analysis and knowledge extraction from their data. The biggest challenge in citizens' engagement proved to be the detailed completion of the diaries, as it was the most time-consuming task.

The assessment of the use of the AQ measuring devices, led to the following results:

- AirVisual Pro was the device with the fewest problems or technical difficulties reported by the citizens. Users had seamless and uninterrupted access to PM concentration levels reported by the device, yet with the drawback of not being able to geo-localise their measurements, due to the absence of a GPS sensor. The online cloud solution for data storage and visualisation proved to be useful for participants allowing them to have easy access to the measurements they conducted.
- Sensebox home was the device for which many citizens reported difficulties in operation, which required that the basic set-up needed to be done by the scientific team supporting the project (i.e. the authors). Moreover, it could not be easily used as a portable device as it required a power bank due to lack of battery and it consisted of individual units wired to each other instead of being a multi-unit, single-enclosure device. On the other hand, the online cloud solution for data storage and visualisation proved to be very helpful in order to obtain the data of

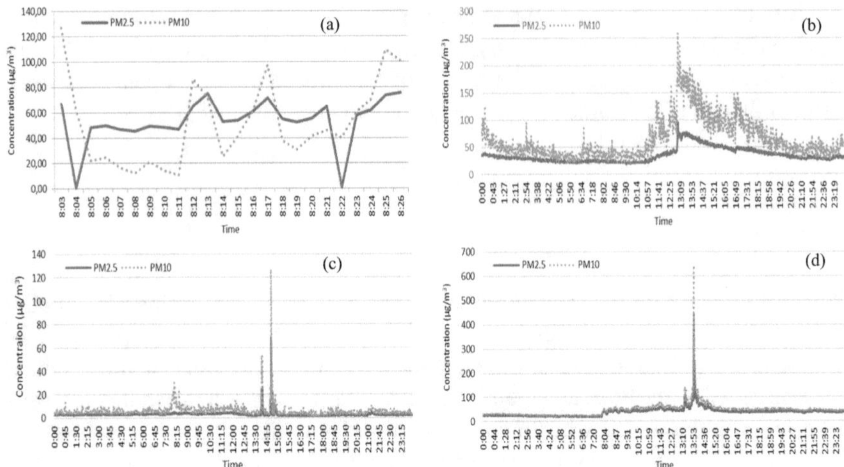

Fig. 1 Typical time-series for each scenario: **a** commuting, **b** accommodation space, **c** my neighbourhood and **d** cooking. Continuous and dotted lines indicate PM2.5 and PM10 concentrations respectively (all values expressed in $\mu g/m^3$). The WHO guidelines for PM2.5 and PM10 are 25 $\mu g/m^3$ and 50 $\mu g/m^3$ mean daily values respectively (www.who.int)

each device and provided the opportunity to supervise the network of devices and to identify problems even before these were realised by users.

- Dylos DC1700 was used by a minority of citizens. Data was stored locally and required processing, as Dylos does not provide concentration of PM, but number of particles which need to be processed in order to obtain PM concentrations. Therefore, data could not be fully reliable because of the assumptions made for the concentrations to be computed.

AQ measurements were conducted by citizens based on the four aforementioned scenarios and were accompanied by event logs recorder in relevant daily reports diaries. Results brought into the foreground interesting relationships between PM10/PM2.5 concentration levels and human activities that affected them. A sample of these results is visualised in Fig. 1.

Figure 1a refers to a small bus, which is reported to be overcrowded and with poor ventilation. Fluctuation of the recorded concentration levels results from the fact that the citizen scientist is sitting next to the bus door, which opens at each stop. The PM2.5 levels are larger than the PM10 levels for some minutes of the studied timeframe, and the overall concentration levels of both pollutants are relatively low in comparison to the WHO guidelines. Figure 1b corresponds to a small apartment with two tenants that stayed indoors for the whole day due to sickness (flu). Interestingly, their presence is linked with increased levels of PM concentrations throughout the day, which exceed the WHO mean daily concentration guidelines for both PM fractions, and for a considerable percentage of the day. In Fig. 1c the participant reported that his/her neighbour is having a barbeque in his yard, which explains the rapid increase of PM concentration and clearly reflects the effect of this activity to AQ. Finally, Fig. 1d corresponds to a citizen cooking (using boiling and baking processes), while the kitchen's ventilator is not operating. Air pollution levels reported by the sensor are particularly high (>600 $\mu g/m^3$) during the time when the cooking activities take place.

In order to extrapolate knowledge from the AQ and diary recordings during the citizen workshop, a two-levelled analysis was conducted by the participants. Initially, they aimed at the identification of pollution sources, so as to designate the main polluting everyday activities, using diagnostic analysis based on time-lapses. In the next step, and having the everyday activities as an input, they targeted at identifying best practices for air pollution reduction via a procedure of a prescriptive analytics type. Citizens appeared to be capable of both pinpointing the main polluting everyday activities and proposing practical measures for lowering either the overall emissions of their impact to the high PM levels, as summarized in Table 1.

Table 1 Citizen workshop results concerning scenarios used, identified polluting everyday activities, and corresponding abatements measures—best practices

Scenario	Polluting activity identified	Proposed air pollution abatement measure
Cooking	• Cooking technique used • Number of people in the room	• Improve ventilation during cooking • Reduce number of people/improve ventilation
Commuting	• High traffic congestion • Overcrowding in buses • Pets in the car	• Commuting with the windows closed • Better cabin ventilation • Alternative route selection • More spacious/more frequent busses • Keeping the pets and the car cabin clean
Accommodation space	• Smoking • Clean by vacuuming • Closed or opened windows • Use of air conditioner • Number of people indoors • Bin full of garbage indoor	• Stop smoking indoors/quit smoking • Improve ventilation when doing household chores • No ventilation when high levels of outdoor air pollution are observed • No ventilation during rush hours when in proximity with high traffic roads • Ventilation when air conditioner is used • Better ventilation when high levels of indoor air pollution are observed • Emptying the bin from the garbage/remove bin from indoor environment

4 Discussion

We used a human-centric approach to study air pollution in relation to everyday human activities. Initially, the main concerns of involved citizens related to air quality were identified and were used as a background for the definition of the AQ monitoring scenarios. Concerning the online citizen communication platform, it was found out that many participants did not make frequent use of it. However, through the online platform many citizens could receive continuous feedback signalling that their work is recognized, while they were able to learn more about air pollution issues and to ameliorate their everyday life, by positively reflecting on their involvement and input to the project. Regarding the latter, we tried not only to offer support to the participants but also to include them in as many project tasks as they desired. We recognized also that some of the participants were getting more engaged to the project as a result of the continuous feedback and information they received by the scientific group.

Results from all four AQ monitoring scenarios provided valuable feedback concerning the causes of high PM concentration in indoor and outdoor activities of everyday life. Participants were able to identify indoor activities directly related to high PM10/PM2.5 emissions and concentrations, and outdoor activities that are also responsible for high PM levels at a personal level, especially while commuting in the city. Through the results of these citizen science activities and with the collaboration of the participants, it was made possible to extract some basic guidelines and good practices to be followed. We were therefore able to determine some basic differences regarding various cooking techniques and inform the participants of how deep-frying for example can result in higher PM concentrations. We were also able to identify the direct relationship between traffic and in-cabin, car or bus high PM concentration levels. Moreover, the number of people and their polluting activities (like smoking, housekeeping etc.) was clearly linked with increased PM levels indoors. Concluding, the results of the AQ measurements provided with insight concerning how some alterations in people's daily routine can reduce exposure to high PM concentrations.

Daily reports (via diaries) proved to be a valuable source of information in order to relate high PM concentrations with certain activities. On the other hand, the level of details and the accuracy of the information provided via the diaries by the citizen scientists varied considerably: detailed daily reports allowed for better correlation between people's activities and air pollution levels. As there were two different types of diaries provided to the citizens (one in digital form and the other in physical), a noticeable difference was identified: citizens with paper-based diaries provided much more detailed and accurate information, compared to participants delivering digital diaries, and the users of the former proved to be more engaged to the project.

The interaction of the research team with participating citizen scientists, helped in accumulating valuable experience with the design of such projects. This led to the re-design of some of the AQ monitoring scenarios in order to make them more personalised and more target-orientated. This interaction also led to a number of suggestions on how to achieve and improve citizen engagement in relevant projects:

- Carefully consider the practicalities of citizen involvement during the design phase of the project
- Secure the necessary human and IT resources that will support interaction with collaborating citizen scientists
- Understand participants motivations
- Manage expectations of participants workload
- Involve and support participants in as many stages as desired
- Establish positive workplace relationship with stakeholders
- Use easily adaptable environmental monitoring hardware and online communication and collaboration tools
- Offer training and learning opportunities to participants
- Address concerns surrounding the quality of data resulting from SC projects
- Provide participants with recognition of their work

5 Conclusions

Most of the participating citizens shared a strong interest on AQ issues and were willing to commit to the agreed agenda, while the deviations are due to two main issues a) difficulties concerning the handling of AQ measuring devices and b) difficulties on filling out the event-driven diaries. Well-structured diaries and easy to install AQ measuring devices are keys issues in such projects.

Results from four AQ monitoring scenarios provided valuable feedback concerning the PM concentration variation related to indoor and outdoor activities of everyday life. Activities which increase exposure to high PM concentration were identified by participants. The majority of these activities were relevant to their everyday utility, such as deep-frying and ventilation in the Cooking scenario, overcrowded means of mass transportation and opening windows while in traffic jam in Commuting scenario, barbeque events in the Neighbourhood scenario and indoor smoking and household cleaning in the Accommodation Space scenario. Proposed best practices were useful but some were abstract, such as ventilation, or ambiguous, such as closing and opening windows in public transportation buses, as these are related with the state of the outdoor air quality, an information that citizens cannot easily acquire in their everyday life. The main aim of the project in terms of AQ was achieved, as citizens were able to relate their daily activities with variations of PM concentration. Concluding, the results of the measurements provided with important insight on how small alterations in daily routine can lead to lower exposure to high PM concentrations when including citizens in the loop of environmental monitoring and relevant knowledge extraction and best practice identification.

Future work will focus on achieving more quantitative results. For this purpose, AQ measurements could also be introduced for additional gaseous pollutants such as ozone or nitrogen oxides. A comparison between URwatair's results and the existing air pollution monitoring stations in Thessaloniki would provide more information on the validity of the low-cost sensors used by URwatair participants. Moreover, it would be interesting to compare the difference concerning the various AQ monitoring scenarios at different seasons of the year. Finally, a nationwide network of air pollution citizen science would add valuable results allowing for a more thorough and quantitative analysis of everyday utility in comparison with air pollution levels indoors as well as outdoors.

Acknowledgements We greatly acknowledge the Green Fund for financially supporting the URwatair project titled "Citizen Participation in air quality and rainwater management in urban areas", (2019–2020), in the frame of the "Innovative actions with citizens" task, "information, awareness raising and citizen training" subtask. We are thankful to the Elliniki Etairia—Society for the Environment and Cultural Heritage, Thessaloniki Branch for coordinating the project as well as to the Hydraulics Division, School of Civil Engineering, AUTh for their collaboration. Special thanks are owed to Mrs Kleopatra Theologidou (project coordinator) and Sotiria Alexiadou (project colleague) as well as to Prof. Kostas Katsifarakis (project scientific advisor) for their support.

References

1. World Health Organization. Ambient (Outdoor) Air Pollution (2018). https://www.who.int/en/news-room/fact-sheets/detail/ambient-(outdoor)-air-quality-and-health. Last accessed 30 October 2020
2. European Environmental Agency: Air quality standards under the Air Quality Directive, and WHO air quality guidelines (2016). https://www.eea.europa.eu/themes/data-and-maps/figures/air-quality-standards-under-the. Last accessed 30 October 2020
3. Zuurbier, M., Hoek, G., Oldenwening, M., Lenters, V., Meliefste, K., Hazel, P., Brunekreef, B.: Commuters' exposure to particulate matter air pollution is affected by mode of transport, fuel type, and route. Environ. Health Perspect. **118**, 783–789 (2010)
4. European Environmental Agency: Report No 19/2019. https://www.eea.europa.eu/publications/assessing-air-quality-through-citizen-science. Last accessed 30 October 2020
5. Schaefer, T., Kieslinger, B., Fabian, C.M: Citizen-based air quality monitoring: the impact on individual citizen scientists and how to leverage the benefits to affect whole regions. Citizen Sci. Theory Practice **5**(1), 6 (2020)
6. Voukantsis, D.; Karatzas, K.; Kukkonen, J.: Intercomparison of air quality data using principal component analysis and forecasting of PM10 and PM2.5 concentrations using artificial neural networks, in Thessaloniki and Helsinki. Sci. Total Environ. **409**, 1266–1276 (2011)
7. Kassomenos, P., Kelessis, A., Petrakakis, M., Zoumakis, N., Christidis, Th., Paschalidou, A.K.: Air quality assessment in a heavily polluted urban Mediterranean environment. Ecol. Ind. **18**, 259–268 (2012)
8. Tolis, I.; Saraga, D.; Lytra, M., et al.: Concentration and chemical composition of PM2.5 for a one-year period at Thessaloniki, Greece: a comparison between city and port. Atmos. Environ. **113**, 197–207 (2015)
9. Popoola, O., Carruthers, D., Lad, Ch., et al.: Use of networks of low cost air quality sensors to quantify air quality in urban settings. Atmos. Environ. **194**, 58–70 (2018)
10. Kumar, P., Morawska, L., Martani, K., et al.: The rise of low-cost sensing for managing air pollution in cities. Environ. Int. **75**, 199–205 (2015)
11. Bhatia, P., Singh, P.: Technological and gamified solutions for pollution control in cognitive cities. In: Ahuja, K., Khosla, A. (eds.) Driving the Development, Management, and Sustainability of Cognitive Cities, pp. 234–249. IGI Global, Hershey (2019)

WISdoM: An Information System for Water Management

Marius Wybrands, Fabian Frohmann, Marcel Andree, and Jorge Marx Gómez

Abstract In the future, equal and universal access to drinking water will become more critical. In this context, collecting, processing, and analysing data is a central part of the strategic decision-making process for water utilities. However, there is a lack of water management information systems that are specifically adapted to the requirements and use cases of water utilities. Therefore, this work presents a prototypically implemented water management information system. The three use cases long-term water demand forecasts, groundwater data management, and precipitation data management were implemented according to the requirements of the water utility Oldenburgisch-Ostfriesicher Wasserverband. This work provides a first software architecture design for a water management information system considering three specific use cases.

Keywords Water management information system · Long-term water demand forecast · Groundwater data management · Precipitation data management

1 Introduction

The climate change and growing exploitation of natural resources calls for an even more responsible and sustainable approach to use non-substitutable water resources [13]. Water utilities provide drinking water for industry and society. Each water utility needs to know about the water quality and quantity in the coming decades, not only

M. Wybrands (✉) · F. Frohmann · M. Andree · J. Marx Gómez
Carl von Ossietzky Universität Oldenburg, Oldenburg, Germany
e-mail: marius.wybrands@uol.de

F. Frohmann
e-mail: fabian.frohmann@uol.de

M. Andree
e-mail: marcel.andree@uol.de

J. Marx Gómez
e-mail: jorge.marx.gomez@uol.de

A. Kamilaris et al. (eds.), *Advances and New Trends in Environmental Informatics*, Progress in IS, https://doi.org/10.1007/978-3-030-61969-5_10

because of sustainable reasons but also due to economic reasons [7]. Knowledge about water quality and quantity is the most important strategic information for various decisions in water management. Information and data about water quality and quantity is used to make investment decisions for the construction of water supply plants [2], to develop adaptation strategies for the effects of climate change [3] and to identify conflicts between stakeholders [24].

However, the management of data and information about water quality and quantity confronts water utilities with challenges [27]: a multitude of influencing factors and interdependencies [24] have to be determined. The spatial transferability of influencing factors is not always possible [6, 27]. Also, the general availability of information is not always given [33]. This leads to the fact that each water utility has established individual processes to monitor and analyse data about the water quality and quantity [15].

The current changes in the availability of government data [21], provides new opportunities for water utilities in different use cases [6]. The current changes enable a critical as well as scientific examination of the technological, process-oriented and methodological possibilities, which may make use of these open data sources to provide new descriptive, diagnostic and predictive data-centered methods [34]. Above all, the integration of heterogeneous external semi and unstructured data into existing structured databases and data warehouses with company-owned data is a challenge [4, 5]. Also, there is a discrepancy between methods that have been tested theoretically and methods that were used in practice [26]. Water utilities often do not have the resources (time and employees) [19] to establish new methods. This leads to workarounds as well as application of inaccurate methods [24].

Therefore, this work describes a software architecture and core functionalities for a water management information system, that primary supports three water management use cases. In the water management information system the three use cases *long-term water demand forecasts*, *groundwater data management*, and *precipitation data management*, are implemented. Data sources of water utility companies (internal) and open data (external) are used to implement the use cases. The water management information system was iteratively developed together with the water utility Oldenburgisch-Ostfriesischer Wasserverband (OOWV) in the WISdoM project over one year. In the first step, workshops with experts were conducted for each use case. In the second step, a use case independent and extensible software architecture was designed and the use cases were prototypically implemented.

In the following, the state of digitalisation of water utilities (Sect. 2), the selection of the use cases (section 3), the software architecture (Sect. 4) and a description of the implemented use cases (Sects. 4.1, 4.2 and 4.3) are described. Finally, an evaluation of the implemented requirements (Sect. 4.4), a conclusion and a outlook is given (Sect. 5).

2 Digitalisation of Water Utilities

Digitalisation potential in the water sector could be summarised under different terms. At the European level, the term "digital water actions" [8] is used. The German Federal Environment Agency uses the terms "water management 4.0" and "water 4.0" [31]. Various research and industry projects (e.g. ICT4Water [9], SWAN [17], and more projects in [31]) focus on the use of digital water technologies. They consider the potentials with regard to the increasing complexity of decision-making. The research in the field covers several technical dimensions like data, methods, use cases and systems.

Wybrands [34] gives an overview of use cases, technologies, and data sources in the context of water management in smart cities. Eggimann et al. [6] and Song et al. [29] discussing opportunities and risks of data-centered concepts based on different data sources. Souza et al. [30] and Sapp et al. [26] presenting the opportunities and risks based on the currently researched methods for linking, processing, and analysing data in water management.

In the case of long-term water demand forecasts, Rinaudo [24], Ghalehkhondaibi et al. [14] and Singh et al. [28] give an overview of methods to forecast the water demand and Liehr et al. [20] describe a process-oriented methodology for long-term water demand forecasting on the example of the water utility Hamburg Wasser. Wybrands and Marx Gómez [33] describing a web-based software prototype that enables water utilities to visualise information about their supply regions. Rueppel et al. [25] present a system for the management of groundwater data and the industrial products AquaInfo of GeoConcept-Systeme GbR and KISTERS Groundwater are specialised in the management of groundwater data. Friese et al. [12] show the current state of research and industry in processing and analysing precipitation data for water utilities.

The research project W-Net 4.0 [10] develops a simulation and data analysing platform, and the project DynaWater 4.0 [10] researches the potential of digital twins in water management. Dmitriyev et al. [5] discussing a software architecture to manage sensor data in water management.

A comprehensive overview of the challenges in German water management is given by the Federal Environment Agency [31]. For classification purposes, Oelmann et al. [23] developed a maturity model for digitisation in water management and applied it to the largest water supply utilities in Germany. Both the Federal Environment Agency [31] and Oelmann et al. [23] conclude that there is a need for practical action for digitisation in water management. In their conclusion, the Federal Environment Agency [31] calls for further development of data generation, storage, and use of data in water management information systems.

3 Selection of Use Cases

The selection of the use cases was carried out in a three-step selection process. Before the project started, the project partner OOWV limited the selection to water supply processes that consider the qualitative and quantitative aspects of drinking water. Business processes related to procurement, controlling, finance, maintenance, or project management were not considered. The area of wastewater treatment was also not considered. The five use cases *laboratory* (qualitative analyses of drinking water), *environmental information act*, *groundwater data management*, *precipitation data management*, and *long-term water demand forecasts* were identified.

In a second step, expert interviews, according to [22] were planned and carried out. Within the scope of these initial expert interviews, the processes were recorded and analysed. The persons, existing data sources and improvement potentials of the existing processes are identified. The expert interviews aimed to gain a comprehensive understanding of the processes. A total of five interviews with experts in the field of *long-term water demand forecasts*, *groundwater data management*, *precipitation data management*, *environmental information act*, and *laboratory* (qualitative analyses of drinking water) were conducted.

In the third step, the results of the expert interviews were processed and discussed. The central part of the preparation was the creation of Business Process Model and Notation diagrams to visualize the considered processes. Personas were created to represent the experts. With these diagrams, the processes of the individual use cases could be compared consistently, which was a useful basis for decision-making when selecting the use cases.

It was decided that the use case *environmental information act* is not suitable as a first use case. Environmental information requests are too complex due to the persons involved, approval processes, departments, data sources, data formats, and return media. The same applies to the *laboratory* use case (qualitative analyses of drinking water). The use cases *groundwater data management*, *precipitation data management*, and *long-term water demand forecasts* could be recorded entirely and were classified as suitable in the discussion. Another deciding aspect was the transferability to other water management processes and the existing processes improvement potentials.

4 Water Management Information System: WISdoM

To address and solve the challenges of water management, a prototype of a water management information system was developed. The water management information system uses a microservice architecture. Figure 1 shows the water management information system architecture and core services. Each use case involves a set of specific microservices that interact and complement each other. Beside the microservices there are six core services (*message broker core service*, *API gateway core*

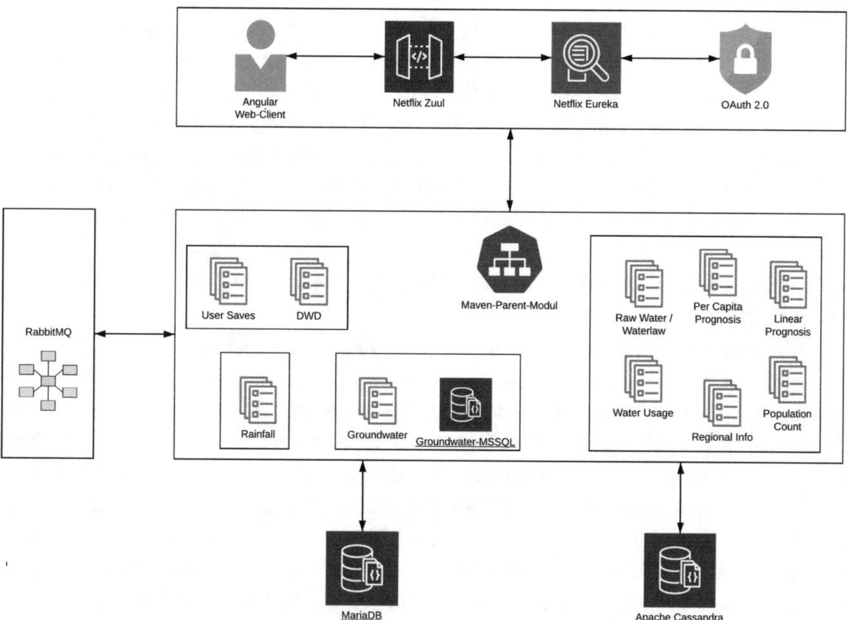

Fig. 1 Architecture of the water management information system

service, *authentication core service*, *load balancer core service*, *service discovery core service* and *web client core service*) that implements management functionalities of the water management information system.

The *message broker core service* provides an integration and communication channel. It is implemented by RabbitMQ. The *message broker core service* ensures that the microservices and core services can communicate with each other.

The *API gateway core service* is the central access point to all core and microservices. The *API gateway core service* is implemented by Netflix Zuul. The *API gateway core service* is supported by a *load balancer core service*, an *authentication core service*, and a *service discovery core service*. The *load balancer core service* and *service discovery core service* are implemented by Netflix Eureka and the *authentication core service* is implemented by OAuth2.0.

The *load balancer core service* is responsible for load distribution if several instances of a particular microservice are initialised or required. When the *API gateway core service* receives a request, it queries the *service discovery core service* to determine a microservice instance. The request is then forwarded to the microservice.

The *authentication core service* is responsible for authenticating incoming requests. It is possible to use different authentication strategies (e.g. local, OAuth, LDAP, Active Directory, OpenID). The implementation takes place using a token-based bearer authorisation using OAuth 2.0. The *authentication core service* uses the Resource Owner Password Credentials Grant, where the user passes his credentials directly to the application [16]. All other microservices and the *web client core ser-*

vice request the *authentication core service* to check the permissions of a user who wants to use the water management information system. In the following paragraphs, the technical core functionalities of the water management information system are discribed and the two core functionalities *metadata management* and *versioning* are discussed in detail. Maven modules were implemented to reduce complexity and allow faster development of new microservices. The modules provide different core functionalities like *REST-APIs*, *database connections* to relational databases or Cassandra, *versioning*, *metadata management*, *personalisable dashboards* and *AMQP-messaging*. Each microservice can add dependencies and use these functionalities in a standardised matter.

Metadata management is one of the technical core functionalities of the water management information system. The aim of having metadata is to attach additional descriptive information to data to ensure transparency of data modifications, processes and analyses. In this context, metadata is descriptive information about individual records. Some examples are the data source, the format of the raw data, etc. The water management information system can add, edit, delete, and retrieve metadata for single rows or even whole data tables. The metadata is saved in an instance of Apache Cassandra due to possible high amounts of metadata. *Metadata management* is implemented in a separate maven module to allow fast implementation in other microservices.

Another technical core functionality is *versioning*. The aim is to allow reproducibility for processes, forecasts, and analyses. Therefore, it is necessary to assign each correction a new version. Old versions still have to be accessible after a new version is created. To achieve this, a new data dimension time was introduced to the data model. The concrete data type is a timestamp. The dimension itself is represented by an extended primary key, that contains the new dimension time. Due to the primary key, each version of a row is now unique per timestamp. This allows querying for all recent data for a given maximum timestamp.

The water management information system uses several internal and external data sources. Internal data sources, for example, are *water consumption data*, *raw water flow rates*, or *precipitation data*. These were made available by the OOWV. An overview of all existing data sources is shown in table 1. Currently implemented external data sources are *population data* and *population trend data* and *weather data* from Germany's National Meteorological Service (Deutscher Wetterdienst (DWD)). Each data source is wrapped by a microservice which allows querying for the wanted data. The data sources can be combined by using the *message broker core service* to receive data from different sources.

A client can access the microservices by sending requests to the *API gateway core service*. The *API gateway core service* queries the *service discovery core service* to get the information on which microservice is accessible on each port. Several microservices are exposing an API. For example, there are analysis and management microservices like *per-capita-prognosis service* or *water-demand-linear-prognosis service*. These need to have access to data sources. Therefore, data is exposed by data microservices, like *water-consumption-data service*, *raw-water-flow-rate-data service*, *water-rights-data service* or *regional-information-data service*. All microser-

Table 1 Overview of available data in the water management information system

Data source	Description	# of rows
Precipitation data	Live data from measuring stations of the OOWV	~770.000
Water consumption data	Water consumption on the basis of individual grid connections and at municipal level (both annual)	~6.500.000
Population data and population trend data	Population figures per municipality, aggregated over one year. Contains forecasts and is based on values by GENESIS Online	~1.700.000
Raw water flow rates	Information about waterworks and their raw water extraction. Aggregated over one year	~600
Regional information	Structure of the OOWV supply areas with allocation of the waterworks	~600
Weather data	Based on data from the Open Data Portal of the DWD	~400.000
Groundwater measuring points	Measured data from groundwater measuring points	~6.000
Water rights	Legally permitted groundwater extraction capacity for waterworks per year	~300

vices can communicate implementing the Advanced Message Queuing Protocol (AMQP).

The user is able to create own dashboards in the web client. He can choose between different data sources. The data can be displayed in graphical form, such as a table or a diagram. In the dashboard, a query builder allows users to perform filtering and aggregation operations without any knowledge of SQL. With this functionality, the water management information system offers a high degree of flexibility in analysis, even for non-technical users.

4.1 Long-Term Water Demand Forecasts

In general, water demand could be defined as the consumption of water measured by the various customers of water utility [24]. Water demand forecasting is a central task of water utilities and pursues several objectives. In addition to operational processes such as short-term expenditure planning [24], they can also support strategic decisions such as infrastructure expansion or adaptation of strategies for the effects of climate change [2, 3]. Furthermore, water utilities are obliged to provide certain stakeholders with the information resulting from the water demand forecasts [1].

However, the preparation of *long-term water demand forecasts* for the next 30 years poses great challenges for water utilities. On the one hand, *long-term water demand forecasting* is a highly interdisciplinary field that has to take into account a multitude of different data sources. Ghalehkhondaibi et al. [14] name historic consumption volumes, climatic variables and socio-economic variables such as population growth rates and economic parameters as relevant variables. On the other hand, these variables are subject to great uncertainty [24]. Besides temporal extrapolation models such as ARIMA, multivariate statistical models, and scenario approaches are mentioned to counter the uncertainties [14, 24].

Moreover, several techniques like artificial neural networks or support vector machines are frequently used for demand forecasting, although they seem to work better for short-term forecasts than for *long-term water demand forecasting* [14]. According to these results, no single method can be considered the best one for *long-term water demand forecasting*. Instead, a hybrid approach could represent a promising option. Besides the selection of suitable forecast models, the integration of significant variables plays an important role as well in improving prediction accuracy. Singh et al. [28] have investigated in further studies that models that only consider proven effective factors perform better than models using all available data. Regarding Ghalehkhondaibi et al. [14], water demand forecasting for different types of consumers and small spatial units such as communes could be a promising approach, where the inclusion of additional variables could lead to more applicable and reliable models.

The research findings are very well aligned with the requirements determined during a workshop with experts from OOWV. Within this creative workshop, a total of 59 requirements, including 123 acceptance criteria, were identified and recorded as story cards. This set of core use case functionalities are summarised in Table 2.

Concerning the research findings, the focus in realising the requirements and core use case functionalities was less on providing a single, perfectly adapted forecast microservice for the OOWV association area. Instead, the use case functionality should be developed, which is going to be flexible and expandable, offering the user various options. The foundation for this flexibility and expandability is the software architecture of the water management information system that is primary based on microservices. Various additional data sources and forecast microservices can be added to the water management information system. In this manner, the *water-consumption-data service*, *raw-water-flow-rate-data service*, and *water-rights-data service* were implemented to make the required internal data sources available for demand forecasting. See Table 1 for data source details.

Furthermore, the integration of external data was illustrated by the development of one additional microservices to provide open access to data sources. Demographic developments received by Genesis Online were encapsulated in a separate microservice that provides historic population censuses as well as different population trend scenarios. These various population trends enable the required implementation of scenario-based approaches considering miscellaneous per capita consumptions. The

Table 2 Overview of the core functionalities for the use case long-term water demand forecasts

Core functionality	Description
Integration of internal and external data sources	Not only internal data sources such as historic *water consumption data*, *raw water water flow rates* and *water rights*, but also freely accessible external data sources such as *population data* or *weather data* should be integrated
Deployment of various forecast services	The water management information system should provide the selection between different forecast microservices. Besides simple mathematical microservices, scenario-based microservices should be available
Selection of spatial units and consumer types	The integrated data as well as the executed *long-term water demand forecasts* should be matched to different selectable consumer types and spatial units such as communes or counties automatically
Expandability	The water management information system should be expandable with further data sources and forecast microservices

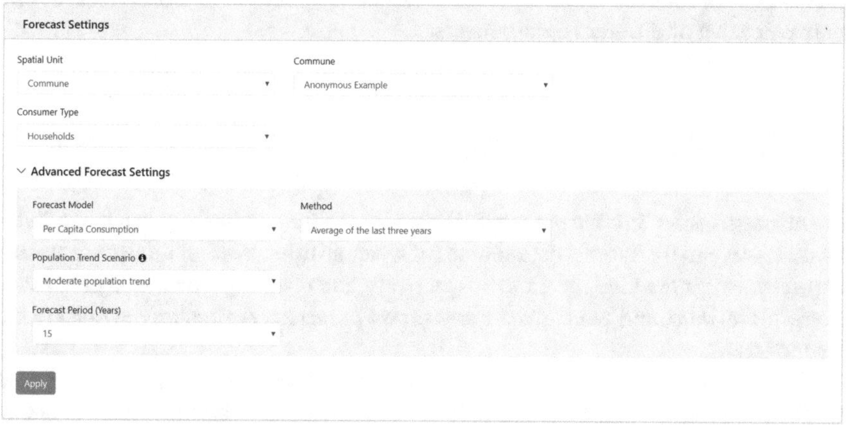

Fig. 2 Water demand prognosis settings

regional-information-data service took on a central role, to overcome the requirement of selecting different spatial units for *long-term water demand forecasting*. For the calculation, the integrated variables must first be converted to the selected spatial unit, such as communes, counties, or waterworks, which is done automatically by an additional microservice.

Also, two exemplary prognosis microservices for executing the actual *long-term water demand forecasts* were implemented. The *water-demand-linear-prognosis ser-*

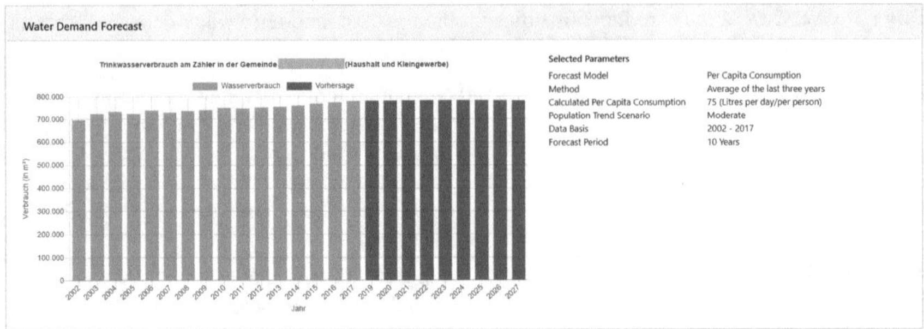

Fig. 3 Result of an executed water demand forecast with selected parameters combining consumption values and population data

vice is a simple regression microservice, which allows a linear as well as a polynomial or logarithmic regression. The per *per-capita-prognosis service* is a scenario-based approach, which calculates *long-term water demand forecasts* based on various population trends in conjunction with automatically calculated or transferred per capita consumptions (see Figs. 2 and 3). According to the implemented microservice architecture, the integrated internal and external data sources, as well as the forecast microservices, could easily be replaced or extended by further microservices to meet more extensive or different requirements.

4.2 Groundwater Monitoring Management

The management of groundwater data illustrates the interaction between the individual use cases and the various data sources of water utilities. This relationship between *long-term water demand forecasts* respectively, the resulting *raw water flow rates*, *precipitation data*, and *groundwater measuring points* are particularly evident in the water cycle.

The European Directive 2000/60/EC describes groundwater as all water, that is below the surface of the ground in the saturation zone and has direct contact with the ground or subsoil. A key objective of the European Directive 2000/60/EC is the establishment of a framework for the protection of groundwater to address both unsustainable water use and water pollution. As a result, European Directive 2000/60/EC requires member states to continuously monitor the quality and quantity of groundwater. Besides chemical pollutants from agriculture and industry, excessive water extraction also represents major environmental pollution [7]. In this context, the quantitative status of groundwater resources must be continually assessed. According to European Directive 2000/60/EC, the status is considered good if the development of groundwater levels shows that long-term groundwater abstraction does not exceed the available groundwater supply.

In Germany, the implementation of this groundwater monitoring is the responsibility of local water utilities. To ensure this, the OOWV has installed about 2 500 groundwater monitoring points in its extraction areas. Most of the measurements are read manually by employees. These and further challenges were identified in a Google Design Sprint [18] with three experts as 55 requirements and 150 acceptance criteria that can essentially be condensed into three use case core functionalities (see Table 3).

Similar to the use case, *long-term water demand forecasts*, the various use case core functionalities for displaying (see Fig. 4) and editing monitoring point data were implemented in form of individual microservices. Further microservices for reporting and managing damaged measuring points were added as well. A central problem with the previous administration and controlling of the hydrographs was the required handling of SQL statements. A generic query builder was developed, which enables the user to generate detailed queries with only limited knowledge and skill in SQL. This query builder allows the employee to create queries without having knowledge of the underlying database structure. The user can select the required operators and variables on the client-side from automatically prepared lists. Furthermore, this fea-

Table 3 Overview of the core functionalities for the use case long-term water demand forecasts

Core functionalities	Description
Overview of the measuring points	All measuring points should be displayed on a map and selectable by the user within this map
Display and adjustment of data	The master data and the groundwater measurements of selected points should be editable. Furthermore, groundwater measurements should be presented as hydrographs which graphically illustrate the observed values over time
Notifications of damage	Damaged or defective measuring points should be reported and managed within the water management information system

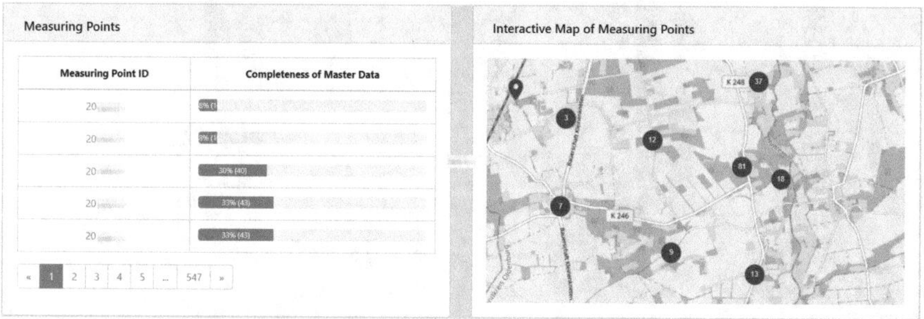

Fig. 4 Overview of groundwater measuring points

ture enables aggregates such as sums, averages, or maximum values of considered variables. Later, this query builder component was adapted and applied in further use cases and data sources.

4.3 Precipitation Data Management

Water utilities require precipitation data for drainage modeling and planning of sewer networks [12]. The relevance of this data is continuously increasing due to the growing number of heavy rainfall events [11]. The public's interest in information is growing. The usage of such data could be helpful for decreasing amount of property damage caused by heavy rainfall event. Therefore, a proper *precipitation data management* is required to process, analyse, and archive the data. The data is also essential for internal planning, as support for operational purposes and communication with the public. A classification of heavy rain events takes place via the specially developed heavy rainfall index (Fig. 5) [12].

In a workshop with two technical experts from the water utility OOWV, 47 functional requirements with a total of 124 acceptance criteria were identified. These

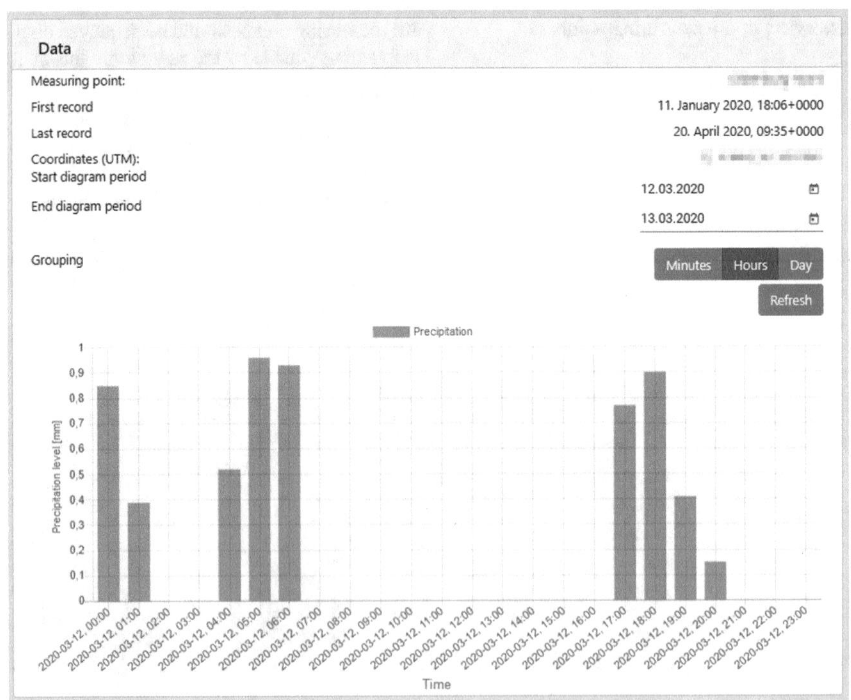

Fig. 5 Visualisation of data gathered by weather stations

Table 4 Overview of the core functionalities for the use case long-term water demand forecasts

Core functionalities	Description
Station data	The user should be able to get an overview of existing precipitation stations (DWD and OOWV), including metadata available to them. The different weather stations (DWD and OOWV) should be visible on a map, and both historical precipitation data and incoming data should be available in minute intervals
Plausibility	The user should have the possibility to create precipitation forecasts based on existing data
Precipitation forecasts	To be able to evaluate existing data in a meaningful way, it should be possible to adjust the resolution the data is displayed. Heavy rainfall events are automatically determined based on selected periods. It should also be possible to compare the data of individual stations
Evaluation	It should be possible to adjust the resolution the data is displayed. Heavy rainfall events are automatically determined based on selected periods. It should also be possible to compare the data of individual stations directly

requirements were described in user stories with a descriptive text, acceptance criteria, and prioritisation. These user stories were then again reviewed by the experts in a second feedback loop. In summary, these 47 requirements can be represented by four use case core functionalities (see Table 4). The data used in the use case are generated from one of the 27 weather stations that measure precipitation and other types of measurements in the OOWV region. Especially the water and sewage plants are equipped with their monitoring stations.

4.4 Evaluation

In order to validate the applicability of the presented approach, a two-stage evaluation was planned in cooperation with the OOWV. First, the usability of the water management information system should be tested by a cognitive walkthrough [32] with expert users. For this purpose, five defined scenarios should be executed by different experts on the water management information system. The comparison of the actual processes with the defined target processes should identify possible improvement potentials. In the second stage, the functionality of the software was to be tested by evaluating the initial requirements in expert interviews with regard to their degree of fulfilment.

Although the flowcharts for the target processes of the cognitive walkthrough as well as the guidelines for the expert interviews have already been prepared, the planned evaluation could not be carried out due to the comprehensive measures in the context of the COVID-19 pandemic in spring 2020. Instead, an internal evaluation was carried out in which all acceptance criteria were assessed on a scale from 1 (not fulfilled) to 4 (requirement exceeded). Of the total of 397 acceptance criteria surveyed, about 46% could not be implemented within the year. However, this high proportion must be put into perspective by two factors. Firstly, the requirements were defined at the beginning of the project without restrictively considering the time frame. Secondly, some requirements of the different use cases were quite similar. These requirements were primarily dealt with in the first use case. In the use case of *long-term water demand forecasts*, therefore, considerably more acceptance criteria could be positively evaluated (62%).

5 Conclusion and Future Work

In this work, a prototypically implementation of a water management information system is presented. The three use cases of *precipitation data management*, *groundwater data management*, and *long-term water demand forecasts* are implemented using a scalable microservice architecture. The microservice architecture enables the integration of external and internal data sources into processes and generates added value for business users. The water management information system can be flexibly adapted to new use cases through core functionalities *REST-APIs*, *database connections* to relational databases or Cassandra, *versioning*, *metadata management*, *personalisable dashboards* and *AMQP-messaging*. By deliberately focusing on only three use cases, a solid software architecture has been developed that provides a foundation for further research in the field of digitisation in water management and water mangemement information systems. The water management information system is currently being extended by the following functionalities:

The *plausibility check* of incoming groundwater measurement data is currently carried out over a value corridor. This will be automated by an anomaly detection using autoencoder. Master and measurement data of assets (water supply plants, sewage treatment plants, pumping stations, etc.) are to be integrated into the water management information system. In the first step, pumping stations with related maintenance processes will be analysed. The *long-term water demand forecasts* are currently using simple regressions. The process of *long-term water demand forecasting* will be extended by new techniques to obtain an even more accurate estimation of future long-term water demand. Extension of the *precipitation data management* to include forecasts provided by the DWD.

Also, some use case core functionalities currently have an idea status, but no practical research is in progress: In addition to quantitative data use cases like *groundwater data management*, *long-term water demand forecasts*, and *precipitation data management*, *laboratory* (qualitative analyses of drinking water) is a high priority use case. These can be integrated into the water management information system. Also, the water management information system can be used to handle complex requests for *environmental information act* requests and to provide corresponding data for citizens. Due to the complexity of the requests for environmental information, it is not possible to limit the requests to particular data sources. The linking of an open data portal to provide information is another idea. In combination with the requests for environmental information as well as the obligatory analyses for waterworks resulting from the national Drinking Water Regulation (TrinkwV), new questions and practical possibilities for research arise. Another idea is the documentation of *water rights cases*. A *water rights case* can be lasting several years with different reports, emails, and analyses. Due to the lack of water management information system, the transfer of the water management information system into an (open source) software product was discussed.

References

1. AG Wasser: Öffentliche wasserversorgung. http://www.ag-wasser.de/wp-content/uploads/2014/01/akwv_daseinsvorsorge.pdf (2002). last access: 28 April 2020
2. Baur, A., Fritsch, P., Hoch, W., Merkl, G., Rautenberg, J., Weiß, M., Wricke, B.: Mutschmann/Stimmelmayr Taschenbuch der Wasserversorgung. Springer Fachmedien Wiesbaden, Wiesbaden (2019)
3. BGS Umwelt: Anpassungsstrategien an Klimatrends und Extremwetter und Maßnahmen für ein nachhaltiges Grundwassermanagement (2010)
4. Blazquez, D., Domenech, J.: Big Data sources and methods for social and economic analyses. Technol. Forecast. Social Change **130**, 99–113 (2018)
5. Dmitriyev, V.,Gómez, J.M., Osmers, M.: Big data inspired water management platform for sensor data. In: INFORMATIK 2015 (2015)
6. Eggimann, S., Mutzner, L., Wani, O., Schneider, M.Y., Spuhler, D., de Vitry, M.M., Beutler, P., Maurer, M.: The potential of knowing more: a review of data-driven urban water management. Environ. Sci. Technol. **51**(5), 2538–2553 (2017)
7. European Commission: A Water Blueprint for Europe (2013)
8. European Commission: Digital Single Market for Water Services Action Plan, pp. 40 (2018)
9. European Commission: Emerging topics and technology roadmap for Information and Communication Technologies for Water Management (2016)
10. Federal Ministry for Education and Research: Project Papers Digital Water Economy—Water 4.0. https://www.fona.de/medien/pdf/Projektblaetter_Digital_Water_Economy_Water_4.0.pdf (2019). Last access: 30 April 2020
11. Field, C.B., Barros, V., Stocker, T.F., Dahe, Q. (eds.): Managing the Risks of Extreme Events and Disasters to Advance Climate Change Adaptation: Special Report of the Intergovernmental Panel on Climate Change. Cambridge University Press, Cambridge (2012)
12. Friese, C., Krämer, S., Bäcker, S.: Zeitlich und räumlich hochaufgelöste Niederschlagsdaten für das Monitoring und die Analyse von Starkregenereignissen. gwf - Wasser|Abwasser **2**, 51–57 (2020)
13. Generalversammlung Vereinte Nationen: General Assembly—Transforming our world: The 2030 Agenda for Sustainable Development (2015)
14. Ghalehkhondabi, I., Ardjmand, E., Young, W.A., Weckman, G.R.: Water demand forecasting. Rev. Soft Comput. Methods **189**(7), 313 (2017)

15. gwf-Wasser|Abwasser. Klimawandel und Wasserversorgung: Baden-Württemberg erarbeitet Masterplan. https://www.gwf-wasser.de/aktuell/02-04-2019-klimawandel-und-wasserversorgung-baden-wuerttemberg-erarbeitet-masterplan/ (2019). Last access 30 April 2020
16. Hardt, D.: The OAuth 2.0 Authorization Framework. RFC 6749 (Proposed Standard), pp. 36–38 (2012)
17. Hauser, A., Foret, N., Combellack, S., Coome, J., Lopez, Q., Hernandez, E., Kharkar, S.M., Rasekh, A., Remy, M., Damour, N.: Communication in Smart Water Networks. Michal Koenig (2016)
18. Knapp, J., Zeratsky, J., Kowitz, B.: Sprint: how to solve big problems and test new ideas in just five days. Simon and Schuster (2016)
19. Levin, E.R., Maddaus, W.O., Sandkulla, N.M., Pohl, H.: Forecasting wholesale demand and conservation savings. J. Am. Water Works Assoc. **98**(2), 102–111 (2006)
20. Liehr, S., Schulz, O., Kluge, T., Sunderer, G., Wackerbauer, J.: Aktualisierung der integrierten wasserbedarfsprognose für hamburg bis zum jahr 2045 teil1: Grundlagen und methodik, pp. 156–165 (2016)
21. Mainka, A., Hartmann, S., Meschede, C., Wolfgang, G.: Stock. Open government: transforming data into value-added city services. In: Foth, M., Brynskov, M., Ojala, T. (eds.) Citizen's Right to the Digital City, pp. 199–214. Springer, Singapore (2015)
22. Mieg, H.A., Näf, M.: Experteninterviews in Den Umwelt- Und Planungswissenschaften: Eine Einführung Und Anleitung. Pabst Science Publications, Lengerich (2006)
23. Oelmann, M., Czichy, C., Merkel, W., Hein, A.: Smart Water Teil 1 Warum die Digitalisierung auch vor der Wasserwirtschaft nicht halt macht (2018)
24. Rinaudo, J.-D.: Long-term water demand forecasting. In: Grafton, Q., Daniell, K.A., Nauges, C., Rinaudo, J.-D., Chan, N.W.W. (eds.) Understanding and Managing Urban Water in Transition. volume 15, pp. 239–268. Springer, Netherlands, Dordrecht (2015)
25. Rueppel, U., Gutzke, T., Petersen, M., Seewald., G.: An internet-based spatial decision support system for environmental data, p. 8 (2004)
26. Sappl, J., Harders, M., Rauch, W.: Maschinelles Lernen in der Siedlungswasserwirtschaft. Österreichische Wasser- und Abfallwirtschaft (2019)
27. Schulz, O., Liehr, S., Grossmann, J.: Fortschreibung und Perspektiven, Das integrierte Prognosemodell für den Wasserbedarf von Hamburg-Szenarien (2017)
28. Singh, G., Goel, A., Choudhary, M.: An inventory of methods and models for domestic water demand forecasting—a review. **35**(3), 12 (2015)
29. Song, M.-L., Fisher, R., Wang, J.-L., Cui, L.-B.: Environmental performance evaluation with big data: Theories and methods. **270**(1-2), 459–472 (2018)
30. Souza, J., Francisco, A., Piekarski, C., Prado, G.: Data mining and machine learning to promote smart cities: a systematic review from 2000 to 2018. **11**(4):1077 (2019)
31. Umweltbundesamt: Chancen und herausforderungen der verknüpfungen der systeme in der wasserwirtschaft (wasser 4.0), p. 142 (2020)
32. Wilson, C.: User interface inspection methods: a user-centered design method. Newnes (2013)
33. Wybrands, M., Marx Gómez J.: Darstellung des ist-zustandes von verwaltungseinheiten am beispiel der langfristigen wasserbedarfsprognose. In: Environmental Informatics: Computational Sustainability: ICT methods to achieve the UN Sustainable Development Goals, vol. 33 of Umweltinformatik, p. 8 (2019)
34. Wybrands, M.: Literaturanalyse von Anwendungsfällen, Technologien und yearnquellen im Kontext Wasserinfrastruktur in Smart Cities. In: Marx Gómez, J., Solsbach, A., Klenke, T., Wohlgemuth, W., (eds.) Smart Cities/Smart Regions—Technische, wirtschaftliche und gesellschaftliche Innovationen, pp. 69–83. Springer Fachmedien Wiesbaden, Wiesbaden (2019)

Exploring Open Data Portals for Geospatial Data Discovery Purposes

Matthias Hinz and Ralf Bill

Abstract Open Data has recently caught much attention from research, politics, and the economy as a resource that fuels innovation and growth. As more data becomes public within an ever-increasing amount of data portals, it can be hard for users of Open Data to keep an overview and find what they are seeking. Therefore, this paper evaluates Open Data portals by their metadata for getting insights about their content and thematic disposition, including measures of interest for the geospatial domain. It summarizes case studies of nine data portals from the German-speaking area of Europe, ranging between national, regional, and local scopes. At these examples, strengths and problems of current data catalogs are shown as well as new perspectives on their content that may improve future data discovery applications.

Keywords Open data · DCAT · Metadata · Geodata

1 Introduction

OpenGeoEdu is a research and development project that fosters the use of Open Geodata, in particular within the German-speaking area of Europe. Former works in this context presented, among others, a web portal [4] that surveys over 388 websites (in July 2020), with open or partially open and cost-free data and services, including 110 Open Data portals, but also geographic, environmental, statistical, scientific, and citizen science data portals. While that survey portal provides a useful entry point for data search, much of its metadata has to be created manually. For instance, the data portals by themselves rarely include a concise abstract of their content that could be automatically obtained and reused, e.g. within the HTML-meta-tags of the website.

M. Hinz (✉) · R. Bill
Professorship for Geodesy and Geoinformatics, University of Rostock,
Justus-von-Liebig-Weg 6, Rostock 18059, Germany
e-mail: matthias.hinz@uni-rostock.de

R. Bill
e-mail: ralf.bill@uni-rostock.de

A. Kamilaris et al. (eds.), *Advances and New Trends in Environmental Informatics*, Progress in IS, https://doi.org/10.1007/978-3-030-61969-5_11

A majority of the portals do not provide machine-readable data at all, especially smaller websites that were not designed with open standards and interoperability in mind. On the other hand, most of the larger Open Data portals in Central Europe offer structured metadata and support at least one of either the DCAT (Data Catalog Vocabulary) standard or the API of CKAN (Comprehensive Knowledge Archive Network), which is one of the most popular software products for Open Data catalogs. Such machine-readable data catalogs enable a decentralized, open and transparent Open Data infrastructure with innovative applications. Using publicly accessible APIs, structured metadata is often retrieved by metadata portals of higher-up institutions, e.g. Germany's national Open Government Data (OGD) portal govdata.de registers datasets from regional or communal data catalogs; the European Data Portal,[1] in turn, includes govdata.de besides 82 other data catalogs from European countries. Many Open Data monitoring tools exist that use this open metadata infrastructure for measuring data availability, quality, openness, and reusability on the country level, as well as per catalog. These tools provide useful insights for policymakers, data experts, and the Open Data community to assure that important data is published in a reusable, timely way and good quality, but they do not necessarily facilitate data discovery. Likewise, this work evaluates catalog metadata, but with a more user-centered approach. A custom parser for CKAN and DCAT interfaces is used to infer information critical for data search purposes, namely, what is the content of a portal, who provides content and how it is provided (i.e. data formats and licenses). Spatial and temporal references of the data are emphasized, as well as data formats and services specific to the geospatial domain. The aim of this work is to use existing metadata for deriving overviews and context of data portals' content by the example of nine case studies. The shown analyses and visualizations might, among others benefits, help users to quickly narrow down their data search and conclude whether or not a portal may contain useful data for their specific purpose. The remaining article is structured as follows. The second section describes related work of Open Data monitoring, evaluation, and discovery. The third section describes the parser and the methodology of information processing. Afterward, selected data portals from Germany, Austria and Switzerland are analyzed: Firstly, the national data portals govdata.de, opendata.swiss and data.gv.at are compared broadly, then portals from federal states and cantons, as well as municipal portals are included in the detailed evaluation. The results are discussed and concluded with further work.

2 Open Data Monitoring and Discovery

Open Data monitoring has been taken on by numerous international organizations, resulting in a variety of reports and web platforms. Especially Open Governmental Data is of high interest since governments of economically leading countries

[1] https://www.europeandataportal.eu.

declared OGD principles as a top priority, prominently in the G8 Open Data Charta of 2013 and its successor, the International Open Data Charta.[2] The state of OGD, mostly focused on country profiles, is monitored by the OECD OURdata Index, the Open Data Barometer by the World Wide Web Foundation (W3C),[3] the Global Open Data Index by the Open Knowledge Foundation,[4] the Open Data Inventory by the non-profit organization Open Data Watch[5] and the European Data Maturity index by the European Union.[6] Academic works in the field of Open Data discovery include Degbelo et al. [3], who surveyed 40 Open Data portals from four European countries. They harmonized and examined terms for data categories, measuring, amongst others, a mean inter-catalog agreement of 0.7 (standard deviation 0.21, range between 0 and 1) for pairwise comparisons of German data catalogs. Based on the extracted common vocabulary, the authors suggested a semantic API for better access to data. The Open Data Monitor[7] is a past project that performed catalog-based monitoring for European countries, where metadata was directly retrieved from major data portals. However, since the project is inactive, the last measurements were performed in 2015, and the numbers of available catalogs and datasets have much increased since then. One of the most extensive and up-to-date monitoring tools for data catalogs at the time of this writing is the Open Data Watch portal and framework [6].[8] It harvests data from 278 data catalogs worldwide and automatically assesses data quality according to 6 dimensions and 19 metrics. For each metric, a score between 0 and 1 is computed. In addition, statistics about data formats, organizations and used licenses are provided. Neumaier et al. also raised concerns about (meta) data quality issues that undermine the searchability of data and the overall success of the Open Data movement. Efficient data search, on the other hand, does not only depend on data quality. As Koesten et al. [5] pointed out, searching data is different from documents because it requires specific skills to download and access data, to interpret data formats and to understand licenses and metadata. Crucial is the context that is given to the data for making sense of it. This context can be derived from the metadata or, as Koesten et al. suggested, from the data itself. The amount of given context depends on the data publishers, but also on the design and user interfaces (UI) of data portals.

[2]https://opendatacharter.net/.

[3]https://opendatabarometer.org.

[4]https://index.okfn.org.

[5]http://odin.opendatawatch.com.

[6]https://www.europeandataportal.eu/en/impact-studies/open-data-maturity.

[7]https://opendatamonitor.eu.

[8]https://data.wu.ac.at/portalwatch.

3 Methods

The following sections describe the technical setup used to obtain and evaluate metadata from Open Data portals, as well as important data properties and formats in this context. Crucial to this work are the methods of data conversion and aggregation that determine the outcomes. Lastly, the concepts of spatial and temporal footprints are introduced, which can be used to derive the spatial and temporal context of a data portal's content.

3.1 Software and Metadata Interfaces

For this work, a custom parser was written in R [10] that sends requests to either DCAT- or CKAN-based interfaces and statistically evaluates the responses. It collects DCAT metadata in RDF/XML format and stores it within a local triple store of the RDF4J open source framework.[9] The stored metadata is then queried using the SPARQL query language and client interface for R [11] and further processed using, among others, the sf package [9] for spatial data handling and ggplot2 [12] for all figures in this paper. CKAN API metadata are fetched as JSON files and processed in R using the package ckanr [1]. A custom R script then translates relevant information to DCAT-compliant RDF-triples and sends them to the RDF4J store, so that analysis can be carried out by similar SPARQL queries as the directly retrieved DCAT data. This translation step is only necessary for the selected Austrian portals. Although a corresponding mapping from DCAT do CKAN was published [2], it was not fully implemented and supported by data.gv.at and the descendant regional/municipal portals at the time of writing. All data were retrieved by end of March 2020. The CKAN API[10] and DCAT[11] rely on different metadata structures, extensively documented online. As previously shown by Neumaier et al. [6], it is yet possible to harmonize the data. For catalogs that support both CKAN and DCAT, the latter was preferred, since it is an official W3C recommendation and part of the European Open Data strategy. The European Commission initiated the DCAT Application Profile (DCAT-AP)[12] as a common specification for data catalogs, content aggregators and data consumers. Consumers, in particular, shall be able to find data more easily through a single point of access.

[9]https://rdf4j.org.

[10]https://docs.ckan.org/en/latest/api/index.html.

[11]https://www.w3.org/TR/vocab-dcat-2/.

[12]https://joinup.ec.europa.eu/solution/dcat-application-profile-data-portals-europe.

Table 1 Metadata terms used for the analysis as defined by DCAT and the CKAN APIs of Austrian governmental data portals

Information	DCAT property	CKAN terms
Dataset (of a catalog)	dcat:dataset	Package
License	dct:license	License
Data format	dct:format	Format
Publisher	dct:publisher	Publisher/Organization
Categories	dcat:theme	Categorization
Keywords	dcat:keyword	Tags
Temporal coverage	dct:temporal	begin_datetime, end_datetime
Spatial coverage	dct:spatial	geographic_bbox, geographic_toponym
Last update	dct:modified	metadata_modified

3.2 Metadata Fields

Metadata fields of interest are shown in Table 1. From this information, different measures can be derived, i.e. the amount of machine-readable data (according to data formats), the amount of spatial data (according to data formats or tags and categories), and spatial and temporal footprints (see below). An important measure that is often invisible to users is the amount of missing or invalid information. Missing information of license, data type or categories greatly reduce chances to find a dataset, as it will not show up when using corresponding filters in the catalog. Also, missing license information creates legal uncertainty for reusing data. Hence, users must be aware of how much they can rely on the metadata.

3.3 Data Aggregation

CKAN and DCAT both distinguish between datasets (respectively packages in CKAN) and resources (respectively distributions in DCAT). A dataset can thus contain multiple resources, i.e. files, services and more. In this work, occurrences of licenses and data formats are only counted once per dataset because the number of files is not a measure for the amount of information. That means, if a dataset consists of multiple files of the same format or license, each unique property is only counted once. Since a dataset may consist of multiple files of different data formats and licenses, the sum of all counts can be higher than the total amount of datasets in a catalog, although this is rarely the case for data licenses. This aggregation practice is also similar to the counts of CKAN catalog UIs and therefore comparable. When counting machine-readable datasets, or spatial data, it is sufficient if one resource in a dataset complies with the criteria, as, for instance, it is common to include reports or documentation in a PDF file to supplement the data.

3.4 Spatial and Temporal Footprints

Machine-readable coverage information for datasets allows the computation of footprints from data catalogs or subset collections of datasets through aggregation. They are represented by lattice data.

In this work, a temporal footprint is a time series where datasets are counted over given time periods, e.g. years. Temporal coverage is given as time periods with begin dates and end dates, hence a dataset is counted if the coverage is overlapping with the given period. As many of the evaluated metadatasets specify a begin date but no end date, it is assumed that such datasets cover any date ahead of the begin date.

Spatial footprints are spatial lattices where datasets are counted over given regions. In this work, a reference region (e.g. country, state/canton or municipality) is taken and divided into grid cells. For each cell, those datasets are counted where the spatial coverage is intersecting it. Spatial coverages are mostly given as a geometry (i.e. polygon) of a geographic bounding box, but may also be exact areas, as well as literal values (text). Since the text descriptions are mostly not standardized and difficult to match unambiguously with regions, they are not used for spatial footprinting herein.

4 Results

The following chapter shows how the described methods are applied to selected data portals and supplies additional context to the results.

4.1 Overview of the Evaluated Portals

For an informative cross-section of the Open Data landscape, Open Data portals for Germany, Austria and Switzerland are examined (see Table 2). The following paragraphs describe the results in comparison of the national, regional (federal states/cantons) and municipal levels. The German national portal has the largest number of datasets, yet, given that Switzerland and Austria are much smaller countries in size and population, the differences are relative. Regional data portals, as for the selected examples of Switzerland and Austria, are often part of the higher-level national portal, yet the German portal datenadler.de (federal state Brandenburg) is an independent platform, though part of the metadata is included in the national portal govdata.de as well. The DCAT-interfaces of data.gv.at and descendant portals for Carinthia and Graz were incomplete at the time of writing, therefore, the above-mentioned CKAN-to-DCAT metadata translation had to be applied.

Table 2 Open data portals and attributes with ISO 3166-1 alpha-2 codes for countries

Portal	Place	Datasets	Metadata
https://data.gv.at	AT (national)	27,202	CKAN,(DCAT)
https://opendata.swiss	CH (national)	7048	CKAN, DCAT
https://govdata.de	DE (national)	38,520	CKAN, DCAT
https://www.data.gv.at Full URL: https://www.data.gv.at/auftritte/?organisation=land-kaernten	Carinthia/AT (federal state)	114	CKAN, (DCAT)
https://datenadler.de	Brandenburg/DE (federal state)	96	DCAT
https://opendata.swiss Full URL: https://opendata.swiss/organization/kanton-thurgau	Thurgau CH (canton)	99	CKAN
https://data.graz.gv.at	Graz/AT (city municipality)	197	CKAN
https://data.stadt-zuerich.ch	Zurich/CH (city municipality)	561	CKAN, DCAT
https://www.opendata-hro.de	Rostock/DE (city municipality)	220	CKAN, DCAT

4.2 National Comparison

Table 3 summarizes the data search features of the national Open Data portals within the countries of interest. Despite relying on similar software and metadata structures, they have major differences, which are also reflected in the metadata contents, as shown below. Notably, only govdata.de offers filtering information by a spatial bounding box on a map or by specifying time intervals. Dataset entries can be sorted in each portal, but the sorting categories and possible directions of sorting (ascending/descending) vary. All portals offer a free text search, but only govdata.de allows for an explicit search within titles and descriptions. Only opendata.swiss offers filtering by political levels (confederation/canton/commune/other), although similar metadata fields (i.e. political geocoding) are specified in the German DCAT application protocol as well.[13] Unlike others, the Swiss portal has no filter for licenses but for terms of use instead, using so-called rights statements.

4.3 Terms of Use

When comparing data licenses and terms of use between the nations (see Fig. 1), many differences become evident. Notably, opendata.swiss, Carinthia and Zurich do

[13]https://www.dcat-ap.de.

Table 3 UI features for data search, sorting, and filtering on national data portals

	data.gv.at	opendata.swiss	govdata.de
List of all publishers	✓	✓	✓ (as a filter)
Sort by	↑↓ most recent ↑↓ Name	↓ relevance ↓ most recent ↑↓ name	↑↓ relevance ↑↓ most recent ↑↓ title
Free text search	✓	✓	✓
Search titles			✓
Search descriptions			✓
Map filter (bounding box)			✓
Time filter (from–till)			✓
Resource type	✓(browse)	✓(browse)	✓(filter)
Political level		✓	
Categories	✓	✓	✓
Tags	✓	✓	✓
Publisher	✓	✓	✓
Responsible body			✓
License	✓		✓
Terms of use		closed/zero / by /restr. commercial use	
License openness			open/restricted
With costs/free	✓ (but all free)		

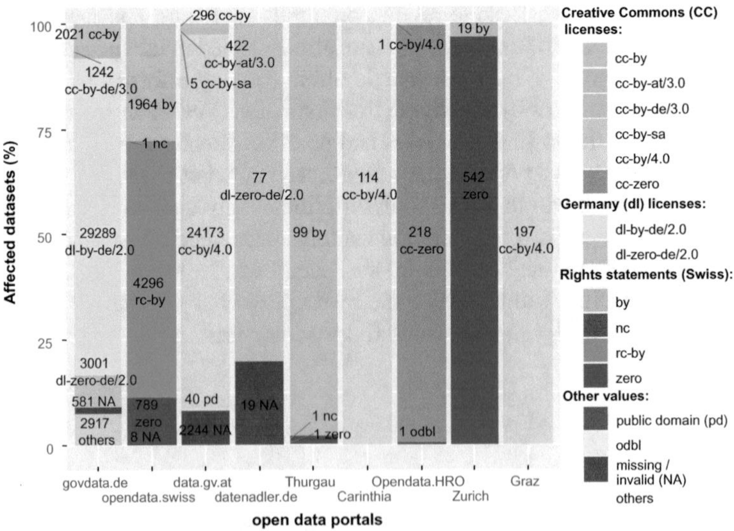

Fig. 1 Terms of use and licenses per dataset for Open Data portals: The bar labels include the count of affected datasets and the relative share of the total count

not use any standard data licenses, except for a minority of 497 irregularly attributed datasets in opendata.swiss. The elsewhere common metadata field *dct:license* is officially not in use by the Swiss DCAT-AP [7] regulation of DCAT. Instead, distributions (resources) are labeled with rights statements (field *dct:rights*) of three elements, specifying if commercial use (1) and non-commercial use (2) is allowed, and whether an attribution of the originator is required (2). For about 39 % of the counted statements in opendata.swiss, any use is permitted with either an attribution requirement (by) or no requirements (zero). One of listed datasets does not permit commercial use (nc). 60.9 % of the statements require attribution and asking the rights holder for permission of commercial use (here abbreviated rc, for restricted commercial use). According to the widely regarded Open Definition [8], datasets that restrict commercial use do not classify as Open Data. However, the portals of Thurgau and Zurich almost completely use *by*- and *zero*-statements respectively. Generally, it can be observed that the distributions of legal terms are homogeneous for the selected regional portals and that they are more heterogeneous on the aggregating national portals. The German and Austrian portals mainly utilize licenses with attribution requirements, although the full legal texts differ. The Austrian portal mainly uses licenses developed by Creative Commons (CC)[14] in different versions. Some datasets use a license that was customized for the Austrian or German jurisdictions (*cc-by-at* and *cc-by-de*), but the majority of CC licenses use the generic legal texts instead, which were drafted by Creative Commons with many jurisdictions and international enforceability in mind. Very few datasets have a share-alike requirement (*sa*) that enforces redistributing the data and derived works under the same or defined compatible licenses. Such a requirement ensures that the data stays *open* for reuse (copyleft) but introduces a slight usage restriction. The German portal govdata.de mainly (86 %) utilizes the data license Germany 2.0 (known as Deutschlandlizenz, abbreviated *dl*, variants *zero* and attribution *by*), which was specifically developed for the German OGD. The federal state-level portal datenadler.de also uses the dl-zero variant, but the municipal portal OpenData.HRO, in contrast, relies completely on the CC-by license 4.0 (except for one OpenStreetMap-based dataset using the Open Database License ODbL). The German data licenses, in contrast to the CC licenses, are much more concise and easier to understand by laypersons, although the CC licenses also provide concise summaries of their full legal text. The bar labeled *others* for govdata.de summarizes 18 different license terms, each covers less than 3 % of the occurrences. That includes 598 datasets released by geodata access laws and 747 labeled as *other-closed*. Missing or invalid license information, in particular for data.gv.at, potentially complicate data reuse and discovery.

---

[14]https://creativecommons.org/licenses.

Table 4 Number of distinct expressions for data formats, for the raw metadata, the harmonized and summarized statistics

	Raw	Harmonized	Summarized
govdata.de	184	115	11
opendata.swiss	51	44	10
data.gv.at	86	67	3
datenadler.de	7	6	5
Thurgau	3	3	3
Carinthia	12	11	6
OpenData.HRO	14	14	10
Zurich	29	26	11
Graz	7	7	6

4.4 Data Formats

Across all selected portals, data formats are difficult to explore or to summarize because many values of the corresponding metadata field are irregular. They often describe the same format with different expressions, e.g. *shape, Shape, SHP, SHAPE-FILE* or use broad terms which are hard to interpret, e.g. *download, diverse, multi* and *www*. High numbers of distinct expressions are shown in Table 4. A coarse harmonization was therefore applied, i.e. by converting all labels to lower case, removing leading and trailing brackets or dots, and mapping (replacing) obviously similar expressions. In a second step less commonly used formats are summed up to one group.

Figure 2 shows the distribution of the most common data formats after the harmonization and summary. The bars labeled as *others* summarize all data formats used for less than 2% of the total count (note: alternatively, percentages of the total amount of datasets could be computed for each group). For govdata.de, this group also includes 2338 datasets where the format information is missing for at least one resource. CSV, a simple data format for tables, is very common on data.gv.at, datenadler.de and the data portal of Graz. Geospatial data formats and services have a greater share within the regional and municipal portals of Carinthia, OpenData.HRO, Zurich and Graz. Although the same data can be found within the respective national portals, starting first from the local portal may therefore be a good search strategy when geodata of a particular region is of interest. For govdata.de, opendata.swiss and the portal of Thurgau, a majority of unspecific or unstructured formats is apparent, such as mere web links (html) or pdf. By browsing through metadatasets of opendata.swiss and Thurgau, it can be observed that many of the resources labeled *service* indeed are WMS or WFS services. However, without further information it cannot be concluded which kind of services are provided. Figure 3 shows a temporal footprint which is computed as described above. The vertical axis spacing is log10-scaled in order to highlight the existence of temporal coverage that include the beginning of the last

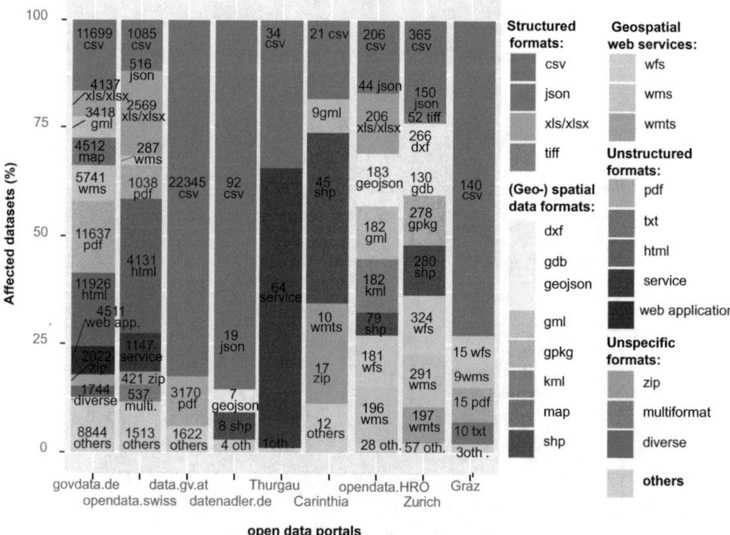

Fig. 2 Distribution of data formats for Open Data portals: The bar labels include the count of affected datasets and the relative share of the total count

century. Each of the portals had the maximum of available datasets between 2014 and 2018 years and a lowering trend for the following time periods (see Table 5). Notably, all data portals, except datenadler.de, provide temporal coverage information for at least some of their datasets.

4.5 *Spatial Footprints*

A spatial footprint with the given methodology could only be computed for govdata.de since for opendata.swiss only irregular literal expressions are given, and for data.gv.at less than 2% of the metadata sheets include geographic bounding boxes for coverage. Spatial footprints of regional portals are less informative. For datenadler.de, only political geocoding references to the state of Brandenburg are provided. Other portals similarly have limited variance and availability of spatial geometries.

The map in Fig 4 highlights the distribution of spatial coverages for govdata.de. While there is no area without coverage by data (minimum: 186 datasets), some areas in northern Germany (around Hamburg) are referred by more than 10,000 sets. The median amount over all grid cells is 1118 datasets.

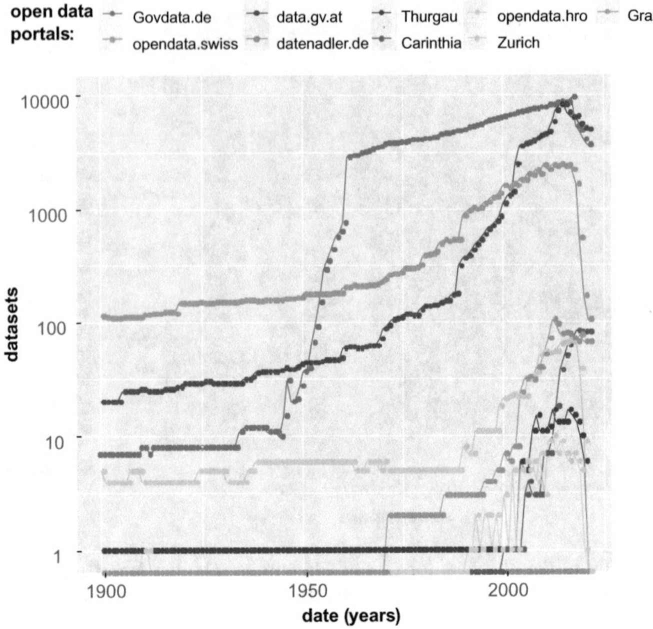

Fig. 3 Temporal footprint with yearly counts of datasets with temporal coverage between 1900 and 2021

Table 5 Maxima of the temporal footprints

	Maximium (year)
govdata.de	9459 (2017/18)
opendata.swiss	2457 (2014/15)
data.gv.at	8204 (2015/16)
datenadler.de	–
Thurgau	17 (2012/13)
Carinthia	84 (2018/19)
OpenData.HRO	10 (2012/13)
Zurich	68 (2016/17)
Graz	106 (2012/13)

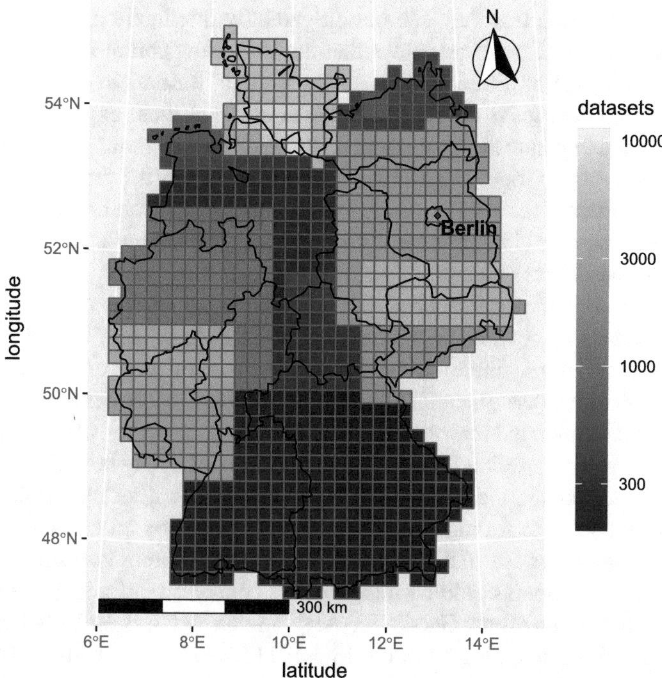

Fig. 4 Spatial footprint of govdata.de with log10-transformed color scale

5 Discussion

One problem of data discovery is the technical and syntactical heterogeneity of the examined data portals and their contents. Although the existing DCAT-based regulations and the common use of CKAN facilitate a certain level of interoperability, the given values in metadata fields are often not standard-compliant or not standardized at all. For instance, the German DCAT-AP.de specifies a set of identifiers for commonly used data licenses, which are often used, but not always. The Austrian portals use different identifiers for the same licenses (although license information is not yet available in DCAT-format) and the Swiss portals make use of rights statements instead. Rights statements, in turn, are denoted by a different metadata field, not commonly used by German and Austrian portals. As current user interfaces mostly reflect on the metadata structures without taking account of the different practices, (i.e. one filter corresponds to one metadata field, resources with missing meta-information for this field are excluded from the results), data search in upstream metadata portals, like europeandataportal.eu, becomes complicated. The computed statistics for this work, in contrast, took account of such irregularities (e.g. consistent labeling and aggregation of licenses as well as rights statements in one Fig. 1 and consistent data formats in Fig. 2 and Table 4), as far as they follow consistent patterns, and therefore demonstrate that they can be mitigated by intelligent application design. More difficulties are caused by values that do not follow common patterns, in particular the multitude of different expressions for data formats in national metadata portals, but also licenses in case of govdata.de. Some of these expressions have the same meaning and just differ in spelling so that they can be homogenized by proper mappings (i.e. dictionaries) or enforced standardization in the process of metadata creation. Other terms (e.g. zip, service) are very unspecific and require revision. As shown above, the overviews are greatly simplified by grouping licenses and format expressions that do not amount to 2–5% of the total distribution (i.e. others). Still, even after a coarse homogenization, there are 8844 occurrences of these less common data formats in govdata.de and more than 1500 in each of the other national portals. For users, it may therefore be difficult to maintain an overview of these formats, even more since many search filters only show the most common terms. Results with rather uncommon but relevant data formats may unknowingly be excluded from search results when applying filters. Hence, further works may be dedicated to categorizing data formats in meaningful terms, for instance, a filter that summarizes all existing geospatial data formats (see Fig. 2), if supported by an extensive mapping, may be more effective for including datasets with less common formats. Many of these metadata problems could be mitigated by consistent enforcement of existing standards and specifications. GeoDCAT-AP,[15] in particular, is an DCAT-AP extension that specifies a mapping from the ISO 19115:2003 core profile and INSPIRE metadata elements to DCAT and RDF expressions. It includes guidelines for spatial and temporal references. Labeling spatial data service types according to GeoDCAT would have greatly improved the information in Fig. 2, where many of those services are just labeled *service*. Missing values are another problem of metadata discovery. As spatial and temporal coverage are mostly treated as optional metadata, and spatial coverage may be a geometry, but often non-standardized placename or description, the available metadata is often not sufficient for generating meaningful footprints. The investigated regional and municipal portals, as far as they provide information, attribute most of their datasets with the same coverage, i.e. referring to the respective federal state, canton or municipality, which makes spatial footprints less informative in regional portals, still useful for aggregating metadata portals on national or global levels. Further works may investigate whether coverage information can be specified more accurately in a pragmatic manner. Also, there are many potential approaches to generate spatial and temporal information not directly available in the metadata. Spatial geometries may be derived from geographic placenames and political geocoding information, provided that they can be unambiguously mapped. For aggregating metadata portals, spatial information could be inferred from the publishers' region of competence or administrative responsibility. Furthermore, missing metadata may be automatically extracted from the referenced datasets themselves, e.g. map services and geospatial data formats often include structured meta-information. Temporal footprints may further incorporate issue and modification dates of metadata. A benefit of footprints over discovery tools in use (see Table 3) is that they provide actual

[15]https://joinup.ec.europa.eu/release/geodcat-ap/101.

visual information of the available data, rather than search filters, which currently just provide a basemap for map search, or no visual information for time filters. Both, spatial and temporal footprints can be generated for fixed time frames or spatial extents, as well es for data subsets, e.g. environmental and geographic information (thematic categories) or any other filter. The aggregation process involved in generating them may be time-consuming though, hence, end-user applications may require a storage of footprints rather than generating them on-the-fly. Contents of the national data portals often change on daily basis, therefore the statistics from March 2020 may be outdated in short time after this writing. Different interpretation of the same metadata may lead to different results, i.e. the spatial and temporal search filters often show a slightly different number of datasets for defined bounding boxes and time intervals. It could be observed that time filter excludes all records which do not have a defined end data in temporal coverage, while herein it is assumed that the temporal coverage expands infinitely to the future. Results must also be carefully interpreted, for instance, a long time-span in temporal coverage may not necessarily refer to historic data. In some cases they refer to zoning plans that were first established decades ago, but the actual resource mostly contains the current version of the plan. Regarding licenses and terms of use, the case studies show that legal practices vary because of country-specific regulations, but also between different portals of the same country. Nevertheless, different terms of use aim to provide similar degrees of freedom for data reuse. Although this work provides some contextual information, discussing the legal implications of these heterogeneous approaches is the subject of extensive juridical debates and therefore out of scope from this work.

6 Conclusions

By investigating 9 different Open Data portals within the German-speaking area of Europe, the given work revealed many faults of the evaluated metadata structures, but also provides insights about the commonalities and differences between the portals. The presented survey and methodology may inspire further efforts to refine data catalog metadata, which could be done by metadata providers and data portal administrators in different stages, including the metadata acquisition within an organization, the enforcement of standards within local and regional portals, and data harmonization within aggregating metadata portals. The revealed need for structural changes may also prompt administrative decision-makers for action. Consistent metadata form the basis of user-friendly data discovery applications, which are either data portals themselves or related services like the portal of data portals from the project OpenGeoEdu. Spatial and temporal footprints are a means to enhance contextual information of a data portal's content, which may help users of Open Data in their data search, provided that they are further developed for an integration in online applications. Highlighting the spatial and temporal context of Open Data catalogs is in the interest of users from environmental and geographic domains but also relevant across other disciplines of data science and technologies.

References

1. Chamberlain, S., Costigan, I., Wu, W., Mayer, F., Gelfand, S.: ckanr: client for the comprehensive knowledge archive network ('CKAN') API (2019). https://CRAN.R-project.org/package=ckanr. R package version 0.4.0
2. Cooperation OGD Österreich: White paper: Ogd metadaten Österreich 2.4. Tech. rep. (2017). https://www.data.gv.at/katalog/dataset/9858d349-7191-467c-8ff0-f1d6cea6f1c2
3. Degbelo, A., Trilles, S., Kray, C., Bhattacharya, D., Schiestel, N., Wissing, J., Granell, C.: Designing semantic application programming interfaces for open government data. eJournal eDemocracy Open Govern. **8**(2) (2016)
4. Hinz, M., Bill, R.: Mapping the landscape of Open Geodata. In: Geospatial Technologies for All : Short Papers, Posters and Poster Abstracts of the 21th AGILE Conference on Geographic Information Science (2018)
5. Koesten, L., Singh, J.: Searching data portals-more complex than we thought? In: CHIIR 2017 Workshop on Supporting Complex Search Tasks, pp. 25–28 (2017)
6. Neumaier, S., Umbrich, J., Polleres, A.: Automated quality assessment of metadata across open data portals. J Data Inform. Qual. (JDIQ) **8**(1), 1–29 (2016)
7. Open Government Data Switzerland: DCAT-AP for switzerland format (2016). https://handbook.opendata.swiss/en/library/ch-dcat-ap
8. Open Knowledge Foundation: The open definition 2.1 (2015). https://opendefinition.org/od/2.1/en/
9. Pebesma, E.: Simple features for R: standardized support for spatial vector data. R Jou **10**(1), 439–446 (2018). https://doi.org/10.32614/RJ-2018-009
10. R Core Team: R: a language and environment for statistical computing. R Foundation for Statistical Computing, Vienna, Austria (2020). https://www.R-project.org/
11. van Hage, W.R., Kauppinen, T., Graeler, B., Davis, C., Hoeksema, J., Ruttenberg, A., Bahls., D.: SPARQL: SPARQL client (2013). https://CRAN.R-project.org/package=SPARQL. R package version 1.16
12. Wickham, H.: ggplot2: Elegant Graphics for Data Analysis. Springer, New York (2016). https://ggplot2.tidyverse.org

Urban Environments and Systems

Developing a Configuration System for a Simulation Game in the Domain of Urban CO_2 Emissions Reduction

Sarah Zurmühle, João S.V. Gonçalves, Patrick Wäger, Andreas Gerber, and Lorenz M. Hilty

Abstract In order to help decision-makers find ways to reduce CO_2 emissions of Swiss cities, a simulation game is being developed within the "Post-fossil cities" project. During the game, participants take on different roles in which they together explore pathways to a future, post-fossil city. An important requirement to the software system of the game was to be easily configurable in order to keep the game adaptive to different target groups of players. We describe a User Interface Management System (UIMS) that has been designed and implemented to realise the flexibility demanded from the game designers' side. The system allows game facilitators to configure the game and decide what kinds of visualisations are used during game sessions. The paper describes how the configuration system was conceptualised, implemented and integrated into the overall system architecture of the simulation game.

Keywords Simulation game · User interface management system · Configuration system · Greenhouse-gas emissions · Sustainable development

---

S. Zurmühle (✉) · J. S.V. Gonçalves · L. M. Hilty
University of Zurich, Zurich, Switzerland
e-mail: sarah.zurmuehle2@uzh.ch

J. S.V. Gonçalves
e-mail: goncalves@ifi.uzh.ch

J. S.V. Gonçalves · P. Wäger · A. Gerber · L. M. Hilty
Empa Swiss Federal Laboratories for Materials Science and Technology, Dübendorf, Switzerland

A. Kamilaris et al. (eds.), *Advances and New Trends in Environmental Informatics*, Progress in IS, https://doi.org/10.1007/978-3-030-61969-5_12

1 Introduction

The climate crisis is a major issue of this century. The domain of urban development offers the potential for substantial reductions in CO_2 emissions and can therefore contribute to reducing global warming. To support decision-makers, a simulation game is being developed within the "Post-fossil cities" project,[1] focusing on sustainable urban development in Switzerland. The players are given the opportunity to simulate the development of a fictional Swiss city and thereby explore different pathways to a post-fossil future. The project, which belongs to the Swiss National Research Program "Sustainable Economy", is positioned in the context of the Paris Agreement [16] and the UN Agenda 2030 for Sustainable Development [15]. In order to achieve the goals of reaching a post-fossil future and of complying with the criteria of a good life, the players interact and cooperate in different roles.

Besides the role-play part, the simulation game includes a physical game board equipped with sensors, a gameplay system that contains a set of simulation models, a simulation system controlling them and a Graphical User Interface (GUI) for the gameplay system. During the game, the players use the game board to manifest decisions, which are registered and evaluated by the gameplay system. Relevant information is visualised to the players on the GUI. These visualisations are composed of different visual components. The simulation game presents visual components on a screen to inform the players about the game status. Different players will find different visualisation types more or less effective. Furthermore, different visualisations can get different messages across. Thus, additionally, a second GUI is needed for the people who configure the software system before a game starts, i.e., the game facilitators. Through this GUI, they can adapt the software to the requirements of the actual game setup, which may vary in the number and type of participants, learning goals, etc.

The applied configurations must be communicated to the gameplay system, which then shows the selected visual components during the following game sessions. The simulation game uses simulation models of different types. The models take parameters, external variables and initial values for internal variables as input and produce time series as output, which can be used in the visual components. Thus, it was required to define a dynamic extraction mechanism for the available visual components in order to flexibly use them in the simulation game system. If the design of visual components changes, the simulation game should be able to easily adapt to those new changes.

Therefore, in addition to the mentioned parts of the gameplay system, a configuration system is needed. This system receives the input of the game facilitators, processes it and sends it to the gameplay system. This configuration system should provide a visual interface that enables the game facilitators to decide which visual components and decision cards (playcards used during the game, sensed by the gameboard) can be used during the game session. It should also define dynamic data extraction mechanisms used to communicate with the simulation system in order to

[1] http://www.nfp73.ch/en/projects/cities-mobility/post-fossil-cities.

obtain the needed model output data. Furthermore, a storage for visual components should be defined and used by the configuration system in order to dynamically load the needed visual components into the system. The flexibility and configurability of the gameplay system's GUI should not be unnecessarily restricted by the interface of the configuration system.

To meet these requirements, we decided to design a User Interface Management System (UIMS). UIMS separate the visual GUI components from the application's logic, which not only reduces a systems complexity, but also allows these two components to be modified separately from each other [12]. Therefore, it is easier to design the application's logic without having to think about the GUI design. A system with a clear structure furthermore facilitates the construction of a robust and adaptable system which is flexible to changes. This paper presents a UIMS in the form of a configuration system for the game developed in the Post-fossil cities project. In order to meet the requirements mentioned above, the UIMS must provide flexible and dynamic interfaces to all connected system parts. Those interfaces should allow the GUI of the game players to remain configurable[2] and flexible[3] with regard to future project requirement changes. This leads to the following research questions:

RQ1: What is a possible structure of a configuration system that allows the gameplay GUI to remain configurable and flexible to simulation game requirement changes?
RQ2: What is a feasible approach to link data streams of an exchangeable backend system to interchangeable visual components, without generating code dependencies in the simulation game software system?

The remaining part of this paper is structured as follows: Sect. 2 presents background information about simulation games and the Post-fossil cities project. Section 3 introduces and discusses related work. Section 4 elaborates on the approaches we used to solve the problem. In Sect. 5, the results are discussed. Section 6 draws a conclusion and presents some suggestions for future work.

2 Background

This section provides some background about the simulation game developed in the Post-fossil cities project and the concept of UIMS.

The Post-Fossil Cities Project's Simulation Game The goal of the "Post-fossil cities" simulation game is to allow stakeholders involved in the development of urban systems to explore possible pathways towards the post-fossil Swiss City in a playful-but-serious manner, while trying to stay within the remaining carbon budget

[2] The term "configurable" applies to the gameplay system. It means that the gameplay system should not be restricted in changing its gameplay mechanics in the future.

[3] The term "flexible" addresses requirements to the gameplay system. It means that the gameplay system should not be restricted in adapting to evolving project requirements.

and at the same time complying with criteria of a good life. The target audience of the game are decision-makers who are committed to a climate-neutral future and students as future decision-makers. The game is developed in an interdisciplinary consortium of researchers and designers from the Technology and Society Laboratory of Empa,[4] the Informatics and Sustainability Research Group at University of Zurich,[5] the Department of Energy and Process Engineering of the Norwegian University of Science and Technology,[6] UCS Ulrich Creative Simulations GmbH[7] and the Institute for Building and Environment of the University of Applied Sciences Rapperswil.[8] The simulation game provides the players with the opportunity to simulate the future development of a fictional city that represents Switzerland. Through interaction, the players try to reach goals related to the UN Agenda 2030 for Sustainable Development [15] and the specifications of the Paris Agreement [16]. By playing the game, participants with different backgrounds get the opportunity to take on new perspectives and to learn about sustainable development in the context of Switzerland. The simulation game allows the players to experiment with different strategies in a "safe" environment. During the game, players take decisions that are evaluated with simulation models, such as a dynamic stock-and-flow model of the Swiss societal metabolism that accounts for the most relevant materials and energy forms. The players get immediate feedback on the impact of their decisions, which allows them to assess their decisions and possibly learn from "mistakes" [18].

Major areas where players can take decisions during the simulation game include the building, transport and energy sectors. The building sector, for example, includes technology-related decisions such as "more frequent renovations with higher standards", "replacement of existing buildings with climate-friendly buildings", "adaptation of heating systems" or "construction of buildings with carbon capture and storage", but also includes lifestyle-related decisions such as "reduction of living space". In a similar way, the transport sector includes decisions focusing on the electrification of mobility and the decisions in the energy sector concentrate on different sources of energy. The thereby created decision space allows players to understand the impact of measures on greenhouse gas emissions as well as the relevant delays involved and thus to understand the importance of timing and sequencing of measures.

The simulation game consists—inter alia—of a physical game board and a software system that contains a heterogeneous set of simulation models. The software system is connected with the game board to be able to read the game status provided by sensors in the game board. The simulation models are used to simulate into the future based on the general development and on the decisions taken by the players. In order to display the results and for other game-relevant visualisations, the system

[4] https://www.empa.ch/tsl.

[5] https://www.ifi.uzh.ch/isr.

[6] https://www.ntnu.edu/.

[7] http://www.ucs.ch/.

[8] http://www.ibu.hsr.ch/.

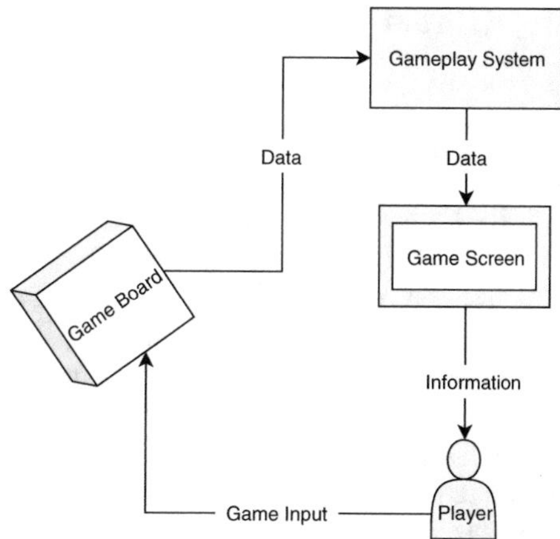

Fig. 1 Interaction between player, game board, gameplay system and game screen GUI

has a GUI. The players interact with the board and the GUI while playing the game. This structure is shown in Fig. 1.

User Interface Management System (UIMS) It is typical for larger software projects that many components have to interact with each other. For instance, if a GUI exists for an application, the input given by the users to the GUI has to be collected, processed and actions in response to the user's input have to be computed. One possible approach to design such a system is to combine the GUI with the mechanism responsible for user input and the application logic methods. However, systems integrating application mechanism and logic with GUI or User Interface (UI) elements have a major drawback: because representation and implementation are tightly coupled, the system is not flexible with regard to later changes. For instance, graphical elements have to be implemented into the application's source-code which increases code complexity—which in turn results in more bugs or more difficult software testing. Thus, having all the different components integrated into the application code makes it hard to maintain the system flexible while in use of GUI applications [12].

To tackle this problem UIMS were developed. The aim of a UIMS is to separate the logical system from the GUI code in a computer program and to maintain communication between them [12]. UIMS allow fast design prototype creation (even before application code is written), enable interfaces to be flexible to changes during development, allow people with different roles in a development team to work together and reduce development time [11]. Hill [5] stated that the usefulness of UIMS is no longer questionable. However, there are needed features that allow UIMS to unleash their full potential. Hill [5] proposed a list of desirable UIMS features that focus on the UI part: Allow the usage of multiple input devices simultaneously, enhance communication between UI modules at run-time and allow the UI to be halted during execution to implement changes. These features may enhance the quality of the UI that the UIMS is managing. While designing a new UIMS, these features can be used as a guideline. Another example of a UIMS guideline can be found elsewhere [10].

3 Related Work

This section presents relevant related work in both UIMS and simulation game research fields.

UIMS There are many implementations of UIMS with several different purposes. Generally, they can be categorised into research-based and commercial products [6]. ProcSee is an example for a commercial product and focuses on the implementation of dynamic GUI [7]. On the other hand, Serpent is one example of a research-based UIMS product. It was developed at the Carnegie Mellon University and enhances the incremental development of UI throughout the whole project phase. It provides a layout editor which can be used for interactive prototyping and a dynamic specification language used for production and also maintenance. Because the architecture is designed in a general way, new interface features could be added even during the development cycle [14].

This paper focuses on the construction of a UIMS in the context of a simulation game and one of the project goals was the construction of a system that lets the GUI of the players be flexible and configurable. Therefore, a general architecture is needed. However, the use-case of this paper is simulation games and not UI development. Therefore, the Serpent's system could not be used for the configuration system UIMS because simulation games depend on more components than the development of UI.

After 1990, more enhanced UIMS were built. One of these UIMS was Alpha [8] which is based on the ideas of the UIMS Serpent. It was designed to make a fast implementation possible which was furthermore clean and powerful. Additionally, Alpha's developers tried to correct the deficiencies of Serpent and its successors [8]. In the research of Shaer [13], the author worked on building dynamically adapted reality-based interfaces by combining a User Interface Description Language (UIDL) and a UIMS to make the system flexible to all kinds of input and output-devices. The UIDL was used to describe and implement the interfaces and keeping the system open to various devices [13]. These kind of researches addressed the issue that modern systems have to be adaptable to all kinds of input and output-devices. In this paper, the configuration system needs to be kept as flexible as possible. Therefore, making it adaptable to multiple different devices is required. However, because all the systems above do not feature simulation games, they could not be used for this paper.

Simulation Games Simulation games are applied in different settings and for different purposes. A number of such games have been applied in the context of sustainable development, amongst others to address environmental issues. For example, Van Pelt et al. [17] investigated the role of simulation games in the context of the communication of climate change uncertainties. They created a simulation game called "SustainableDelta" which was used in a workshop with students and water managers. Like in the "Post-fossil cities" simulation game, the players got the chance to discuss decisions and related consequences [17]. Another example for a simulation game is Septris [4], an online mobile simulation game in the context of health that features the detection and treatment of sepsis.

4 Developing the Configuration System

The infrastructure of the "Post-fossil cities" simulation game, excluding the configuration system developed here, consists of an interactive game board, a gameplay GUI application displaying information about the current game state, models used to run simulation experiments (e.g. a stock-and-flow model to calculate stocks and flows of materials and energy in Switzerland) and an agent-based simulation system which handles all the tasks necessary to orchestrate the simulation.

The game board is a tangible object which lets the players physically interact with it by playing cards. It includes sensors that recognise the players' actions mediated by so called decision cards via sensors and process them directly or send them to the game's backend system for further computation. For example, a decision card could state that the player made an investment in solar energy. Playing this card triggers an update of the current state of the game, to which the players will react again.

The gameplay GUI on a screen is designed to present visualisations in order to inform all players about the game actions and states. Those visualisations are composed of visual components. They can be common charts such as line charts or more creative ones like a filling glass of water that represents a carbon budget. During the game, the visualisations are updated according to the played decision cards of the players. Thus, the players always see the impact of their actions. The gameplay GUI's coupled backend system manages which visual components and information are shown and how they behave on the display during the game. The gameplay backend gets its data from the simulation system, which selects the appropriate models to use. For instance, a model could calculate data about yearly CO_2 emissions. The output of those models are used as data source for the visual components shown on the gameplay GUI and the configuration system's GUI. The UIMS interacts with the gameplay's backend system and the simulation system.

The configuration system handles three interfaces: the configuration data transaction to the gameplay system, the model's data extraction from the simulation system and the definition of the visual component storage. The transaction of the configuration data is done by transferring a JavaScript Object Notation (JSON) file containing the data. The configuration system communicates with the simulation system via Application Programming Interface (API) calls. The visual components designed for the simulation game are stored in a git repository and are dynamically loaded into the configuration system.

The UIMS consists of a frontend GUI, a backend system, a database and an external data storage. Figure 2 shows an overview of the overall simulation game's structure. All components which belong to the configuration system are indicated with the colour orange. A git repository is used as a data storage for visual components. This part is coloured in purple. The blue components are part of the gameplay system and the green box represents the simulation system. The following sections will elaborate on the separate system parts of the configuration system. The source code of the configuration system can be found in a digital repository [19].

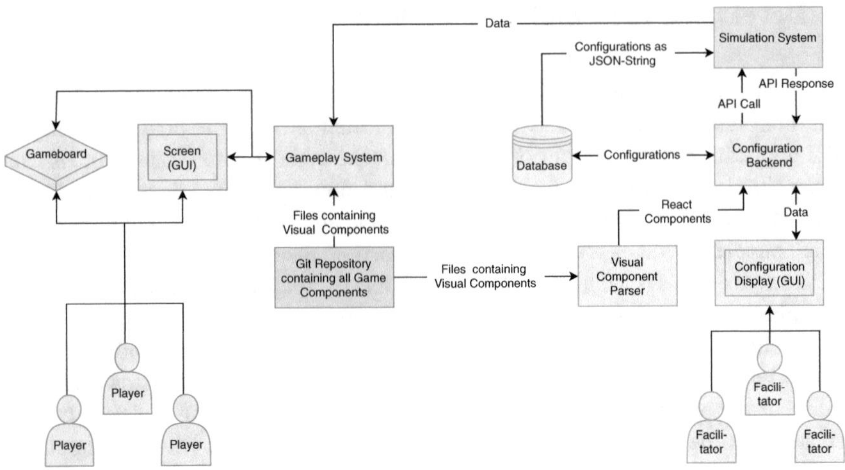

Fig. 2 Simulation game project structure

Frontend The configuration system is a web-application. A web-application is not device bound, which means that every device that has access to the web server of the simulation game has access to the configuration system's application. The frontend part consists of the GUI that the game facilitators interact with to set the configurations. Its most important requirement was to enable the facilitators to make all needed configurations and update the backend system about the made changes. The following were the UIMS frontend's requirements:

1. It must be possible to choose from which git repository to load the visual components from.
2. It should be possible for the facilitators to decide which of those visual components are then shown on the gameplay GUI screen.
3. Facilitators should be able to decide which decision cards can be used during the game session.
4. It should be possible to arrange the chosen visual components on screen.
5. The facilitators should be able to upload their configurations.

These requirements were collected by conducting interviews with the UIMS's stakeholders, i.e. the game's current facilitators. To keep the system as flexible as possible, facilitators must be able to configure from which git repository the visual components are imported from. The storage of the visual components is therefore not limited to one specific git repository. Thus, to fulfil the first requirement, a separate webpage was created which enables the user to enter a link to the git repository of the visual components. A screenshot of this webpage is shown in Fig. 3a. To fulfil the second and third requirement, a separate webpage was created which is split into two parts: the visual components settings, shown in Fig. 3b, and the decision cards settings, presented in Fig. 3c. All available visual components and decision cards are listed in a separate checkbox list from where they can be enabled or disabled. For each

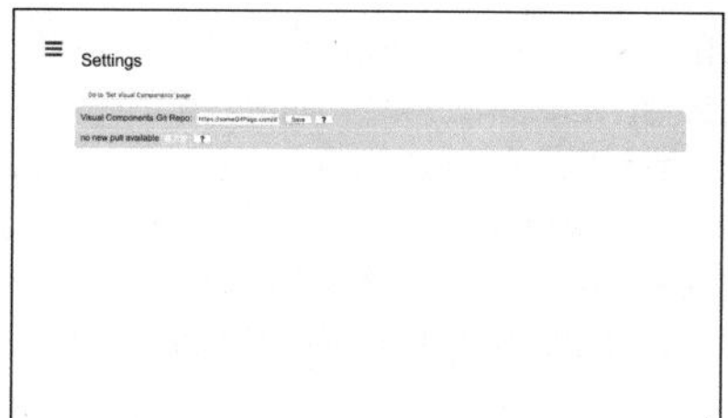

(a) Settings webpage enabling the game facilitator to enter a git repository link

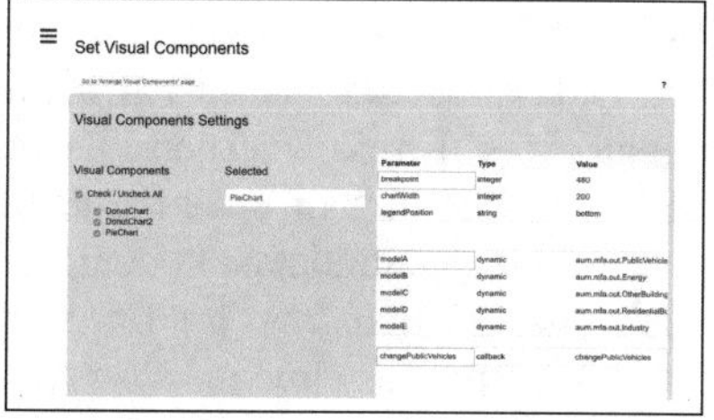

(b) Webpage enabling the game facilitator to decide which visual components should be used during the game session

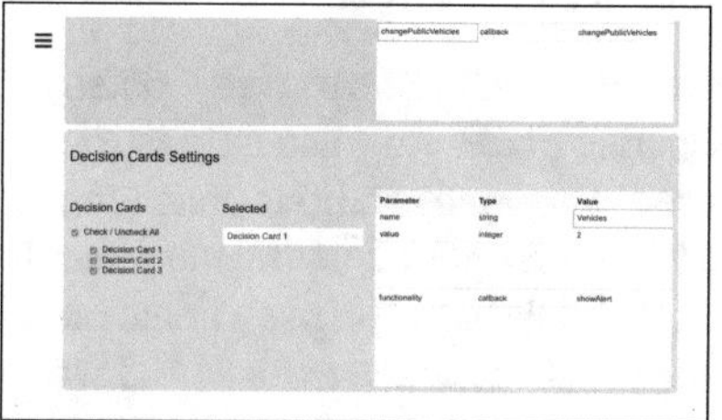

(c) Webpage enabling the game facilitator to decide which decision cards should be used during the game session

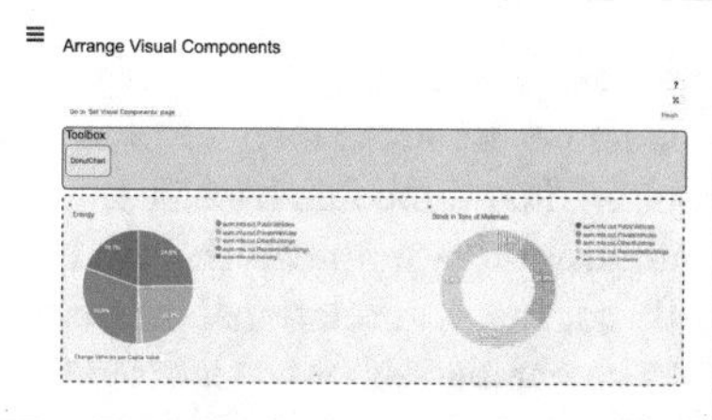

(d) Webpage enabling the game facilitator to arrange visual components on screen - here one of three components is located in the toolbox

Fig. 3 Screenshots of the web-application of the configuration system

visual component or decision card their corresponding parameters can be modified in a separate box. Parameters can be static, e.g. the colour of the visual component, or dynamically linked, e.g. which model in the simulation system provides the data that the visual component needs.[9] Lastly, in order to fulfil the fourth and fifth requirement, a third webpage was designed which focuses on the arrangement of the chosen visual components. A screenshot of this webpage is shown in Fig. 3d. All previously chosen visual components are shown on the screen. If there are too many components at a time on the screen, individual components can be placed within a toolbox for temporary storage. The size of the components can be adjusted and if the game facilitator is satisfied with their arrangement, they can save the settings by pressing a finish button. By pressing on a preview button, the facilitator can see how the arrangement is going to look like on the gameplay system's interface. By using those webpages, the facilitators can modify all needed simulation game settings before each game session starts.

[9] Because the output of a model itself changes dynamically during a game session, these parameters are called dynamic.

Backend While the frontend of the configuration system is responsible for the GUI display, the backend manages all application logic and interface connections of the system. It is therefore responsible for managing the connection to the database (including inserting or updating and extracting information), receiving from and write data to the simulation system, defining API calls which are used to send to and receive information from the frontend system and loading the visual components from a git repository into the system.

The backend system follows the Model-View-Controller (MVC) design pattern [9]. The application logic, e.g. extracting data from a data storage, is performed in the *Model* part of the design pattern. The *Controller* part is responsible for preprocessing the data and defining API calls between the backend and frontend. The *View* is however implemented in the frontend and defines the GUI display.

Database The configurations made with the UIMS's GUI are saved to a database. A database can store data permanently in a separate place. It is also easy to extract required data if the database is properly managed. The configurations done with the configuration system can therefore be stored for a long period of time and can be easily accessed. A relational database was used, which is based on a relational model [2]. The database is only accessed by the backend system, which extracts and writes data into it.

Interface Between UIMS and Gameplay System To communicate the configurations made in the configuration system's web interface to other system parts in the simulation game, a JSON file is used. JSON files have a tree like structure, which allows the information to be easily extracted. It is also quite simple to add additional information, without changing the existing content [3]. Thus, using JSON format to store information makes it easier for other system parts of the simulation game to process the configuration data. Of course, the data can also be extracted from the database directly. However, to do so multiple SQL queries have to be written as the data is stored in multiple tables across the database. This approach would not be flexible to future changes in the database. Furthermore, information in a JSON file can be accessed with one call. Therefore, additionally storing the information in a JSON file to communicate the configuration data keeps the system simpler and more flexible.

Git Repository Data Storage On the configuration webpage, the game facilitators can view all available visual components and arrange them on screen, as to set how the final gameplay GUI should present them. It makes most sense to directly show the visual components as they are going to look like when they are used during the game session. Thus, the configuration system needs access to all visual components's source code in order to visualise them directly on the configuration GUI. Furthermore, the visual components are bound to change, which means that the configuration system should always have access to their updated version. Storing the visual components in an external repository, such as git, makes for a clean solution in the context of the configuration system. Git is a distributed version control system used to store source code and track its development [1]. The features of git are perfect for the configuration system's use-case. Storing the visual components elsewhere and loading them on-demand into the system keeps the system flexible to changes. The designers of the visual components can upload them into a git repository. Whenever the visualisations change or are updated, the designers simply need to push their new version onto the repository. The configuration system on the other hand only needs the URL to the git repository that stores the visual components and, by cloning it, gets immediate access to the source code of all visual components. Their code can then furthermore be used in the UIMS's frontend system to visualise the visual components on screen.

Because hardcoding the link to the git repository into the configuration system is inflexible to changes, the facilitators can update the repository URL by using the system's GUI. Thus, the facilitators do not need to know about how the visual components are stored or updated. They only need to know that they need to provide the link to the right repository to the configuration system. After submitting the link, the frontend system transfers it to the backend system by using API calls. The backend then checks the URL, clones the corresponding git repository and updates the database with the new git repository link. Thus, when the git repository link was once submitted through the configuration GUI, it does not need to be re-entered again. The visual components from the entered git repository are then available for usage in the configuration system.

However, there is one major issue when using a git repository as a visual components storage: How can the configuration system detect if a file on a given git repository contains one or multiple visual components? It is not a flexible approach to assume that all files in the repository only contain one visual component. This assumption restricts the designers of the visual components too much because they would lose the possibility to write multiple classes or methods into one file. Therefore, a simple annotation language was created to detect visual components. These annotations are placed within the documentation strings (docstring) of given objects, such as classes, methods or functions. Furthermore, the annotations define which model data the visual components need. In order to define the model output data, static values or dynamic paths to the location of the datastreams can be used. To read the annotations, a parser was implemented which walks through all files in a given folder and checks if a file contains any visual components. It furthermore extracts all needed information stated in the annotations which is needed to include the visual components in the configuration system.

By using these annotations, all visual components can be integrated into the configuration system. Even when the location of the git repository changes, the process of extracting the visual components stays the same which makes the system flexible to newly defined or updated visual components.

UIMS and Game GUI Integration To integrate the configuration system into the simulation game architecture, the connection between the gameplay GUI and UIMS must be established. The gameplay GUI has to be extended in order to successfully combine those two system parts. The following additional components have to be created: a configuration data extractor, API calls between the gameplay system and the simulation system and a visual components loader as well as parser. In order to get access to the configuration data of the UIMS, the gameplay system has to extract the configuration data from the configuration system's database. Because all configuration data is available as a JSON structure within the database, only this file has to be extracted using a single SQL call. JSON allows easy access to its stored data, therefore extracting the configuration data can be established by using existing methods. The gameplay system then knows which visual components it should visualise, what their size is and where on the screen they are located. The system also knows which data the visual components need in order to be visualised correctly. Static data can directly be used. However, if data paths are given, the system has to extract the data from the given path inside the simulation system. API calls are used to exchange this data. The extracted data has then to be passed to the visual components. In order to extract the source code of the visual components, the gameplay system has to use the same mechanism as in the configuration system to clone the given git repository. The link to the repository is also given in the configuration JSON. Because the visual components contain the annotation language described above, the parser of the configuration system can be used in order to check which files in the repository contain visual components. As a result, the gameplay system knows which classes or methods it has to import in order to visualise the visual components on screen. Because the dynamic data is extracted directly from the simulation system, the visual components data is always up to date. By completing those steps, the configuration system can be integrated with the gameplay GUI of the simulation game. Of course, this procedure is not limited to this specific use-case but can be used for various other simulation games with similar system components.

5 Discussion

Research Question 1 In order to answer the question "*What is a possible structure of a configuration system that allows the gameplay GUI to remain configurable and flexible to simulation game requirement changes?*", the configurations made in the configuration system have to be flexible to changes. The communication channel between the UIMS and the gameplay system consists of a single JSON file. The JSON syntax is in itself flexible to new content because it is structured like a tree. New content can simply be added without changing the whole structure [3]. Therefore, even when the gameplay system changes in structure and requires more information, the JSON output file can still be used because the new content can easily be added. If configuration requirements change in the future, the gameplay system is still able to change its architecture. The JSON output file of the UIMS remains flexible to newly defined requirements as well as to changes in the configuration process. Thus, the gameplay system remains flexible in its definition and structure. Therefore, the answer to RQ1 is to design a UIMS with the properties stated above.

Research Question 2 To answer the question "*What is a feasible approach to link data streams of an exchangeable backend system to interchangeable visual components, without generating code dependencies in the simulation game software system?*", designing a dynamic loading of visual components and their corresponding model data was essential. Given that it is possible that the gameplay system uses other types of visual components than the ones designed, a dynamic visual component extraction mechanism is important for the simulation game. Because the configuration UIMS allows to point to a location where the visual components are defined (in the form of a git repository link), the visual components can simply be swapped out by others. Furthermore, the annotation language enables the system to recognise visual components in a given collection of code files. This annotation language makes it possible to extract all needed information used to load the visual components in the system. Thus, visual components can be dynamically loaded into the system without having to include any additional code in the UIMS's source code. This procedure has the advantage that the gameplay system can use other visual components in the future and the configuration system does not need to be adapted. The game facilitators simply have to input a new link in the settings page of the web-application which loads the new components automatically. Additionally, the defined annotation language and its corresponding parser are not limited to the defined configuration UIMS, but can also be used in other simulation game system parts that use visual components, such as the gameplay system's GUI.

The data used in the visual components are extracted from the simulation system. However, this data processing system might also change. If the gameplay system needs other information for the visual components in the future, the configuration system's structure also does not need to be changed. The only thing that needs to be adapted is the data's location in the visual components annotations and the API between the configuration system and the simulation system. Therefore, the extraction of the model data is not limited to one source which makes the system flexible to changes and provides a dynamic structure. Thus, the answer to RQ2 is to design a UIMS with the interface properties stated above.

6 Conclusion

We showed the design and implementation of a UIMS in the specific form of a configuration system for the simulation game of the Post-fossil cities project. The UIMS enables the gameplay system of the simulation game to remain flexible and configurable even when project requirements evolve in the future. It allows the game facilitators to decide which visual components and decision cards will be used during the game session and to decide which dataset these components will use. The UIMS consists of a web-application with a frontend, backend and database part. To communicate the configurations of the configuration system to the gameplay system, the JSON format is used which has a very flexible structure that is easily extendable. To dynamically load the visual components into the system in order to let the facilitators modify them, the UIMS contains a git loader and parser that clones a given git repository, which contains the visual components, and extracts the corresponding files by using an annotation language. This structure allows the visual components to be independently designed as they do not need to be added directly as source-code into the configuration system. The data used for the visual components is stored externally in the simulation system. Overall, the configuration system lets the game facilitators define where the visual components are stored, allows them to define which components are used and let them specify the used model data. The final configuration is then stored into a database, where it can easily be extracted for further usage in the gameplay system. Due to the flexible structure of the UIMS, the gameplay system's flexibility and configurability is still provided when integrating the UIMS into the overall simulation game architecture of the Post-fossil cities project. However, the configuration system is not limited to be integrated with the presented simulation game, but could also be integrated with other kinds of simulation games. The flexible structure of the configuration system allows such a use-case extension.

In a future version it should be possible to store multiple configurations in the UIMS. Only one configuration can be edited and stored at a time so far. Furthermore, the configuration can only be stored in one language. For the future, it would be beneficial to enable the creation of different language versions of the same configuration. This feature would allow to use the settings made by the game facilitators for different groups of users who have similar interests but speak different languages.

To sum up, by using the configuration system in the simulation game, the game can be adapted to various types of target groups, enabling the game to unleash its fullest potential. By adapting how the simulation game represents information, the simulation game can bring across several messages to the players, making the game experience as effective as possible and helping players finding ways to reduce CO_2 emissions in future urban development.

Acknowledgements This work was supported by the Swiss National Science Foundation (SNSF) within the framework of the National Research Program "Sustainable Economy: resource-friendly, future-oriented, innovative" (NRP 73) Grant-N° 407340_172402/1. Aside from the authors, the following team members were involved: Markus Ulrich, Marta Roca Puigròs and Daniel Müller.

References

1. Chacon, S., Straub, B.: Pro Git, 2nd edn. Apress (2014)
2. Codd, E.F.: A relational model of data for large shared data banks. Commun. ACM **13**(6), 377–387 (1970). https://doi.org/10.1145/362384.362685
3. Crockford, D.: The application/json Media Type for JavaScript Object Notation (JSON). https://tools.ietf.org/html/rfc4627 (2006). Last visited: 13 April 2020
4. Evans, K.H., Daines, W., Tsui, J., Strehlow, M., Maggio, P., Shieh, L.: Septris: a novel, mobile, online, simulation game that improves sepsis recognition and management. Acad. Med. **90**(2), 180 (2015)
5. Hill, R.D.: Some important features and issues in user interface management systems. SIGGRAPH Comput. Graph. **21**(2), 116–120 (1987). https://doi.org/10.1145/24919.24928

6. Iannella, R.: A graphical user interface reference model for messaging systems with directory integration. Ph.D. thesis, Bond University (1994)
7. IFE Institute for Energy Technology: Procsee graphical user management system technical overview. https://ife.no/wp-content/uploads/2018/12/ProcseeTechOverview.pdf (2018). Last visited 13 April 2020
8. Klein, D.: Developing applications with a uims. In: Proceedings of USENIX Applications Development Symposium, pp. 37–56 (1994)
9. Krasner, G.E., Pope, S.T., et al.: A description of the model-view-controller user interface paradigm in the smalltalk-80 system. J. Object Orient. Program. **1**(3), 26–49 (1988)
10. Lane, T.G.: A design space and design rules for user interface software architecture. Technical report , Software Engineering Institute, Carnegie-Mellon University, Pittsburgh, PA (1990)
11. Myers, B.A.: Creating User Interfaces by Demonstration. Academic Press Professional Inc., San Diego (1988)
12. Olsen, D.: User Interface Management Systems: Models and Algorithms. Morgan Kaufmann Publishers Inc., San Francisco (1992)
13. Shaer, O.: A framework for building reality-based interfaces for wireless-grid applications. In: CHI '05 Extended Abstracts on Human Factors in Computing Systems (CHI EA'05), pp. 1128–1129. ACM, New York, NY, USA (2005). https://doi.org/10.1145/1056808.1056845
14. Software Engineering Institute: Serpent overview. Tech. Rep. CMU/SEI-89-UG-2, Carnegie Mellon University (1989)
15. UN General Assembly: Transforming our world: The 2030 agenda for sustainable development (2015). A/RES/70/1
16. UNFCCC: Adoption of the Paris agreement. In: United Nations Framework Convention on Climate Change. Report No. FCCC/CP/2015/L.9/Rev.1 (2015)
17. Van Pelt, S., Haasnoot, M., Arts, B., Ludwig, F., Swart, R., Biesbroek, R.: Communicating climate (change) uncertainties: simulation games as boundary objects. Environ. Sci. Policy **45**, 41–52 (2015)
18. Wäger, P.: Post-fossil cities. http://www.nfp73.ch/en/projects/cities-mobility/post-fossil-cities (2018). Last visited 13 April 2020
19. Zurmühle, S.: isr-ifi/pfc-uims: Pfc-uims v1.0.0 (2020). https://doi.org/10.5281/zenodo.3690735

Estimating Spatiotemporal Distribution of Moving People in Urban Areas Using Population Statistics of Mobile Phone Users

Toshihiro Osaragi

Abstract This paper proposed a method for estimating the spatiotemporal distribution of static and transient populations of urban areas by using population statistics created from the location information for users of cell phones. The advantages and disadvantages of the various population statistics available were evaluated and methods were investigated for integrating the data while using their strengths to best advantage and compensating for weaknesses.

Keywords Spatiotemporal distribution · Mobile spatial statistics · Konzatsu-tokei® · Person trip survey · Moving people

1 Introduction

There is a growing demand for data that allows highly accurate understanding of the spatiotemporal distribution of both moving and static people in urban areas. In disaster mitigation planning, for instance, it is essential to have information about the spatiotemporal distribution of transient occupants in a city. However, the transient occupants and passengers in a large metropolitan area change dynamically, which suggests large potential variations in the number of individuals affected by an earthquake [1–4]. Currently, a variety of population data is available (Table 1), but none of it provides an accurate understanding of numbers and departure/arrival points of moving people using detailed units of space and time.

Person Trip survey data (PT data) focus on people's spatial movement. PT data are based on responses to questionnaire surveys and provide much information, including the sex, age classification, purpose of movement/stay, means of transportation, departure/arrival locations and times, etc. There are a large body of researches using PT data [1–7]. These studies were all attempts to compensate for the shortcomings of PT data and provide useful background for this study, which has the same theme. However,

T. Osaragi (✉)
School of Environment and Society, Tokyo Institute of Technology, Tokyo, Japan
e-mail: osaragi.t.aa@m.titech.ac.jp

A. Kamilaris et al. (eds.), *Advances and New Trends in Environmental Informatics*, Progress in IS, https://doi.org/10.1007/978-3-030-61969-5_13

Table 1 Characteristics of population statistics

Population statistics	Target	Spatial resolution	Temporal resolution	Survey frequency	Attribute info
Census data	Residence-based population	250 m grid [all Japan]	–	Every 5 years	Age, gender, residence, family members
Person Trip Survey (PT data) [1–7]	Travel behavior of persons aged 5+	Municipal zone	Every minute on a weekday	Every 10 years	Departure/arrival locations and times
Mobile Spatial Statistics (MSS) [8–14]	Mobile phone users (sample rate: ca. 40%)	500 m grid (all Japan)	Every 1 h	Everyday	Age, gender, residence
Konzatsu-toke$^{i®}$ [15, 16]	GPS data (sample rate: ca. 0.5%)	500 m grid (all Japan)	Every 5 min	Everyday	Residence, occupation
Agoop [17–19]	GPS data	Point data (all Japan)	Every 1 h	Everyday	Direction and velocity of movement

the PT data were taken at 10-year intervals, so they do not help in overcoming the lack of fresh data.

Mobile Spatial Statistics (MSS) from mobile phones provide regional populations in grid-cell units at any desired time, which are provided by NTT DoCoMo Inc. These are population statistical data, the number of cell phones using the cellular network, and incorporate the penetration ratio among cell phones operated by DoCoMo [8–14].

Konzatsu-tokei® refers to people flows data collected by individual location data sent from mobile phone under users' consent, through Applications provided by NTT DOCOMO, INC. Those data are processed collectively and statistically in order to conceal the private information. Original location data is GPS data (latitude, longitude) sent in about every minimum period of 5 min and does not include the information to specify individual. Some applications such as "docomo map navi" service (map navi/local guide) [15]. However, these data are also obscured by the process of anonymization [16]. In this paper, by integrating multiple sets of data, including PT data, MSS, and Konzatsu-tokei®, we examined a method of estimating the spatiotemporal distribution of moving and static people in urban areas, and construct a method of estimating the numbers of people flowing in/out of each grid-cell per unit time and the spatiotemporal distribution of moving people.

2 Method of Estimating the Number of People Flowing in/out of Each Grid-Cell Per Unit Time

The variables used in the mathematical descriptions are shown in Table 2. This paper employed MSS data, which consists of detailed records featuring high sampling fractions and precise temporal and spatial units, as the basic information expressing the spatiotemporal population distribution. However, population M_i^t (Figs. 1A) occupying cell i at time t, which is obtained from the MSS, fails to distinguish between transient and static occupants. Therefore, we attempted to make this distinction by defining the fraction of static occupants (static occupant fraction $_aP_i^t$, Figs. 1B) obtained from the PT data using the following method. M_i^t, obtained from the MSS, was imposed as a constrain and a maximum likelihood algorithm was written to hold the sum of the number of static, entering, and exiting occupants to M_i^t. This provided estimates for $_an_i^t$, the number of static occupants, $_bn_i^t$, the number of exiting occupants, and $_cn_i^t$, the number of entering occupants (Fig. 2).

Next, data for spatial movements were compiled from Konzatsu-tokei®, which contains detailed spatiotemporal information about departure/arrival locations and times. The fraction of population movements between cells i and j during time span t to $t + \Delta t$ is denoted p_{ij}^t (Fig. 1C$_2$).

Last, the maximum likelihood estimator T_{ij}^t for the number of individuals moving between grid-cells i and j is found via the inter-grid-cell motion fraction p_{ij}^t, using the number of individuals leaving grid-cell i, $_bn_i^t$, and the number of individuals entering grid-cell i, $_cn_i^t$, in the criterion.

When the populations M_i^t and $M_i^{t+\Delta t}$ and the static occupant fractions $_aP_i^t$ and $_aP_i^{t+\Delta t}$ are known, the following equations for the maximum likelihood estimator for the number of static occupants satisfying the criteria can be derived:

$$_an_i^t = \frac{\left(M_i^t + M_i^{t+\Delta t}\right) - \sqrt{\left(M_i^t + M_i^{t+\Delta t}\right) - 4PM_i^t M_i^{t+\Delta t}}}{2P} \tag{1}$$

Table 2 Variables used in mathematical description

Variables	Description
$_an_i^t$, $_aP_i^t$	Number of people and population fraction occupying grid-cell i
$_bn_i^t$, $_bP_i^t$	Number of people and population fraction exiting grid-cell i
$_cn_i^t$, $_cP_i^t$	Number of people and population fraction entering grid-cell i
M_i^t	Population occupying cell i at time t, which is obtained from the Mobile Spatial Statistics (MSS)
p_{ij}^t	The fraction of population movement between cells i and j during time span t to $t + \Delta t$
T_{ij}^t	The maximum likelihood estimator for the number of individuals moving between grid-cells i and j

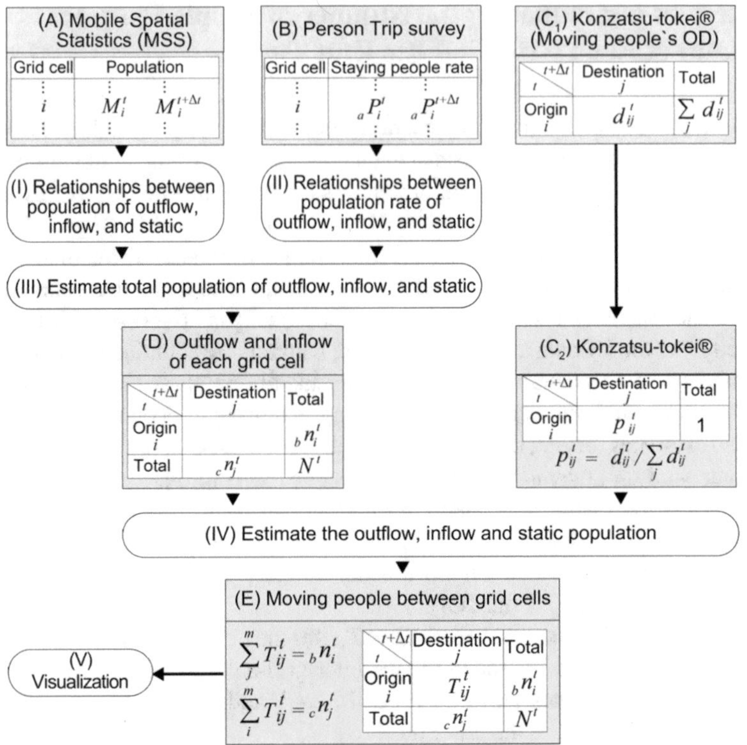

Fig. 1 Integration method for multiple demographic datasets

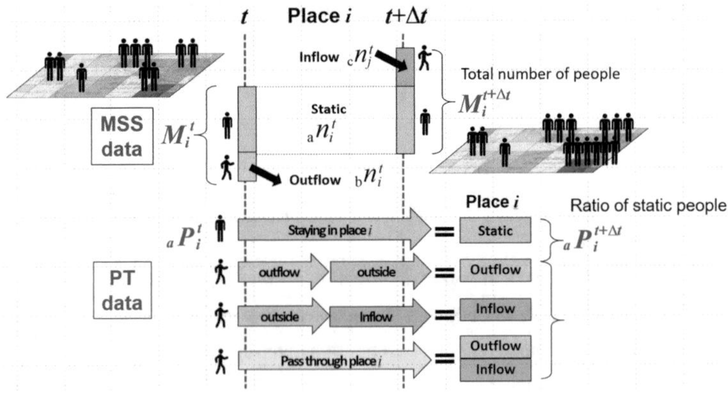

Fig. 2 Relationships between outflow, inflow, and static people

where $P = \frac{{}_aP_i^t + {}_aP_i^{t+\Delta t} - 1}{{}_aP_i^t \, {}_aP_i^{t+\Delta t}}$.

Once the number of static occupants ${}_an_i{}^t$ in grid-cell i during time span t to $t + \Delta t$ has been found, the number of people exiting grid-cell i ${}_bn_i{}^t$ and the number of people entering grid-cell i ${}_cn_i{}^t$ can be calculated.

The numbers of people leaving (${}_bn_i{}^t$) and entering (${}_cn_i{}^t$) are called "moving population data" and are used in criteria for the hourly calculations of population distribution. Additionally, the inter-grid-cell motion fractions $p_{ij}{}^t$ between grid-cells i and j during time span t to $t + \Delta t$ can be calculated from Konzatsu-tokei®. These are used to calculate the maximum likelihood estimator for individuals moving between grid-cells i and j during time span t to $t + \Delta t$. Between the numbers of people leaving and entering and the number of individuals moving between grid-cells $T_{ij}{}^t$, we establish equations shown in Fig. 1E. (Note that m denotes the number of grid-cells.)

The number of individuals moving between grid-cells $T_{ij}{}^t$ is calculated as follows, employing the inter-grid-cell motion fractions $p_{ij}{}^t$ obtained from the Konzatsu-tokei® under the above the maximum likelihood estimators providing the highest values for the occurrence probabilities.

$$T_{ij}^t = p_{ij}^t \times A_i^t \times B_j^t, \tag{2}$$

where $A_i^t = \frac{{}_bn_i^t}{\sum_i^m p_{ij}^t B_j^t}$, $B_j^t = \frac{{}_cn_j^i}{\sum_i^m p_{ij}^t A_j^t}$.

Variables $A_i{}^t$ and $B_j{}^t$ are mutually dependent, but arbitrary starting values are chosen for a converging calculation, and this will provide the unique value for the number of individuals moving between grid-cells $T_{ij}{}^t$.

3 Method of Estimating the Number of People Moving Between Grid-Cells Per Unit Time

In the Konzatsu-tokei®, the number of people moving between grid-cell i and grid-cell j in a 5-min period is obtained as an aggregated value every 60 min. Based on this, the movement probability between grid-cells i and j in a 5-min period from time t, $p^t{}_{ij}$, is found (Fig. 1C$_2$). The movement probability in unit time Δt is found as element of matrix $[\boldsymbol{p}^t]^{\Delta t/5}$, which is matrix $\boldsymbol{p}^t$ containing the element $p^t{}_{ij}$ raised to the power of $\Delta t/5$ (Fig. 3).

The number of people moving between grid-cells i and j between time t and $t + \Delta t$, $T_{ij}{}^{t\text{–}t+\Delta t}$, is estimated by a maximum likelihood estimation using the two matrices of the number of people flowing out ${}_bn^t{}_i$, the number of people flowing in ${}_cn^t{}_i$, and the movement probability $[\boldsymbol{p}^t]^{\Delta t/5}$, found by the above procedure.

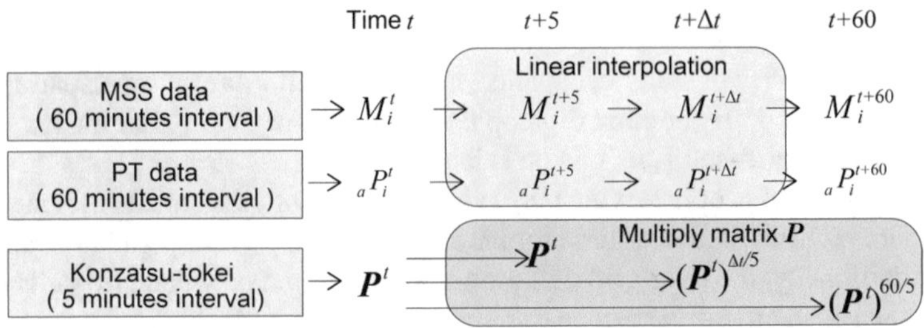

Fig. 3 Unifying time interval of datasets

4 Temporal Change in Number of People Flowing in/out of Each Grid-Cell

The number of moving people in 500 m grid-cell units is estimated using data for the range and dates/times shown in Fig. 4.

Figure 5 shows temporal changes in 60-min units in the number of people in three specific grid-cells, as well as the number of people flowing in/out of these three grid-cells. The figure shows that, at Shinagawa Station (Tokyo city commercial/business district), in the morning, the number of people flowing in is higher because people are commuting to work, but during the day, the numbers of people flowing in/out are low and people stay within the area, while in the evening, the number of people flowing out increases due to people returning home (Fig. 5a). Meanwhile, at Ojima (Tokyo residential suburb), in the morning, there are many people flowing out and the number of people in the grid-cell decreases, while in the evening, the number of people flowing in gradually increases due to people returning home and the number of people in the grid-cell increases (Fig. 5b). Furthermore, at Shinjuku Station (large-scale transfer point), in the morning, the trends are similar to the commercial/business

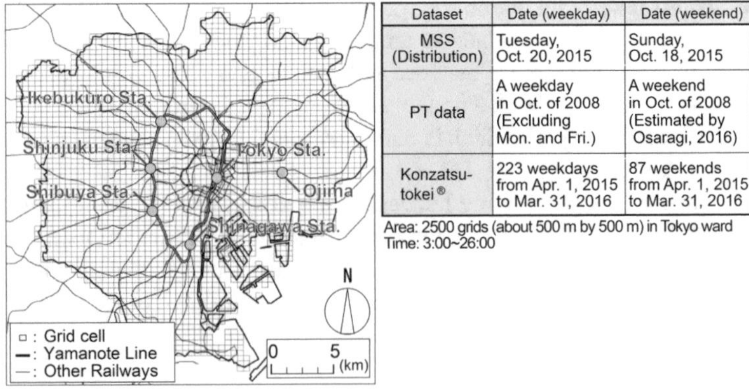

Dataset	Date (weekday)	Date (weekend)
MSS (Distribution)	Tuesday, Oct. 20, 2015	Sunday, Oct. 18, 2015
PT data	A weekday in Oct. of 2008 (Excluding Mon. and Fri.)	A weekend in Oct. of 2008 (Estimated by Osaragi, 2016)
Konzatsu-tokei®	223 weekdays from Apr. 1, 2015 to Mar. 31, 2016	87 weekends from Apr. 1, 2015 to Mar. 31, 2016

Area: 2500 grids (about 500 m by 500 m) in Tokyo ward
Time: 3:00~26:00

Fig. 4 Study area and data used for analysis

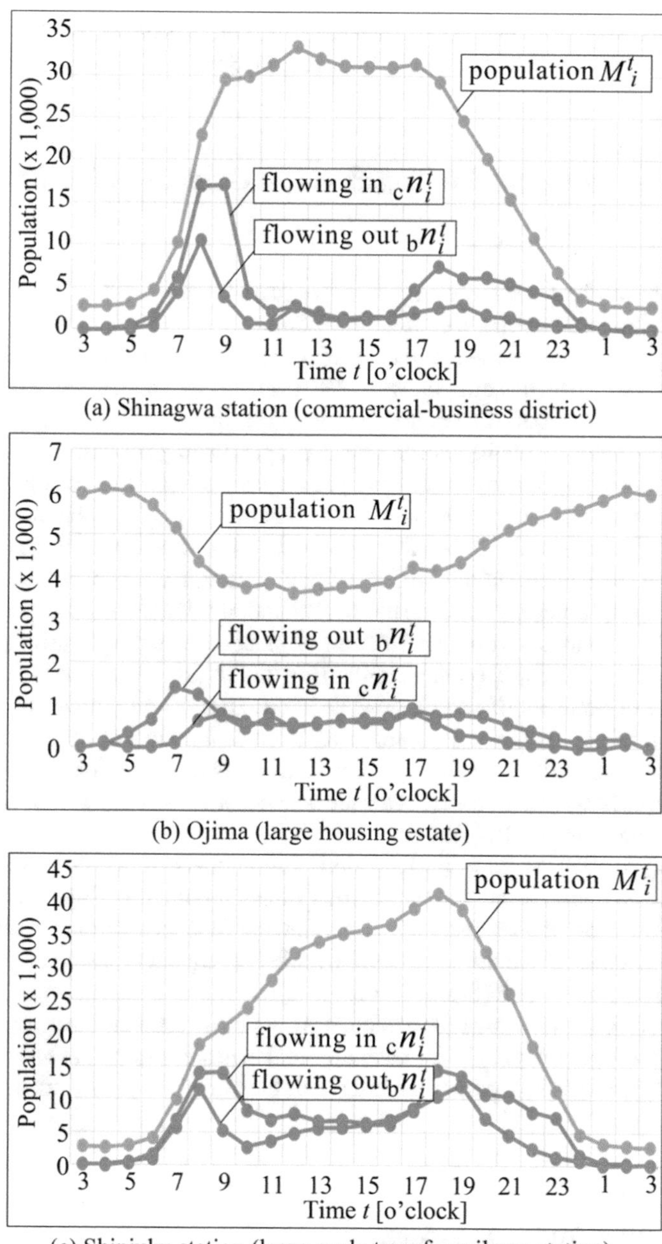

Fig. 5 Time fluctuation of the number of outflow and inflow people

district, but during the day and in the evening, the numbers of people flowing in/out are high and the turnover of people is intense (Fig. 5c).

5 Number of People Moving Between Grid-Cells Per Unit Time

Figure 6 shows the temporal changes in the spatial distribution of the number of people flowing in/out during a certain time period.

By integrating multiple sets of data, it is possible to distinguish between moving and static people, which is difficult with MSS (Fig. 6a), and to determine the number of people flowing in/out of each grid-cell per unit time (5, 60 min, etc.) (Fig. 6b). Additionally, focusing on the distribution of people flowing in/out of a specific grid-cell, we estimated the number of moving people per 5-min/60-min periods, using the number of people flowing in/out and the movement probability obtained from the number of moving people of the Konzatsu-tokei®. The results are shown in Fig. 6c.

Focusing on Shinagawa Station, in a 5-min period, there is a difference in the numbers of people flowing in/out on the Tokyo Station side and the Kanagawa Prefecture side, but the trends in the distribution of places where people are flowing out to and flowing in from are roughly similar (Fig. 6c(1), (2)). However, looking at the distribution of numbers of people flowing in/out after 60 min, it can be seen that the numbers of people flowing out are lower than the numbers of people flowing in. Furthermore, the outflow destinations are concentrated on the Tokyo Station side, but the people flowing in come from a wide area and are not limited to the Kanagawa Prefecture side (Fig. 6c(3), (4)).

At Shinjuku Station, in a 5-min period, the numbers of people flowing in/out from inside and outside the Yamanote Loop Line are both high (Fig. 7(1), (2)), but in a 60-min period, the numbers of people flowing in from a wide area outside the Yamanote Loop Line and the numbers of people flowing out to inside the Yamanote Loop Line tend to be high (Fig. 7(3), (4)).

As described above, using this method, it is possible to estimate the spatiotemporal distribution of moving people for each arrival/departure grid-cell at any unit time, and to analyze trends in departure/arrival points and direction of movement of people flowing in/out.

6 Summary and Conclusions

Ordinal modeling techniques of environmental informatics based only on static physical objects like buildings do not always describe a certain aspect of their realities.

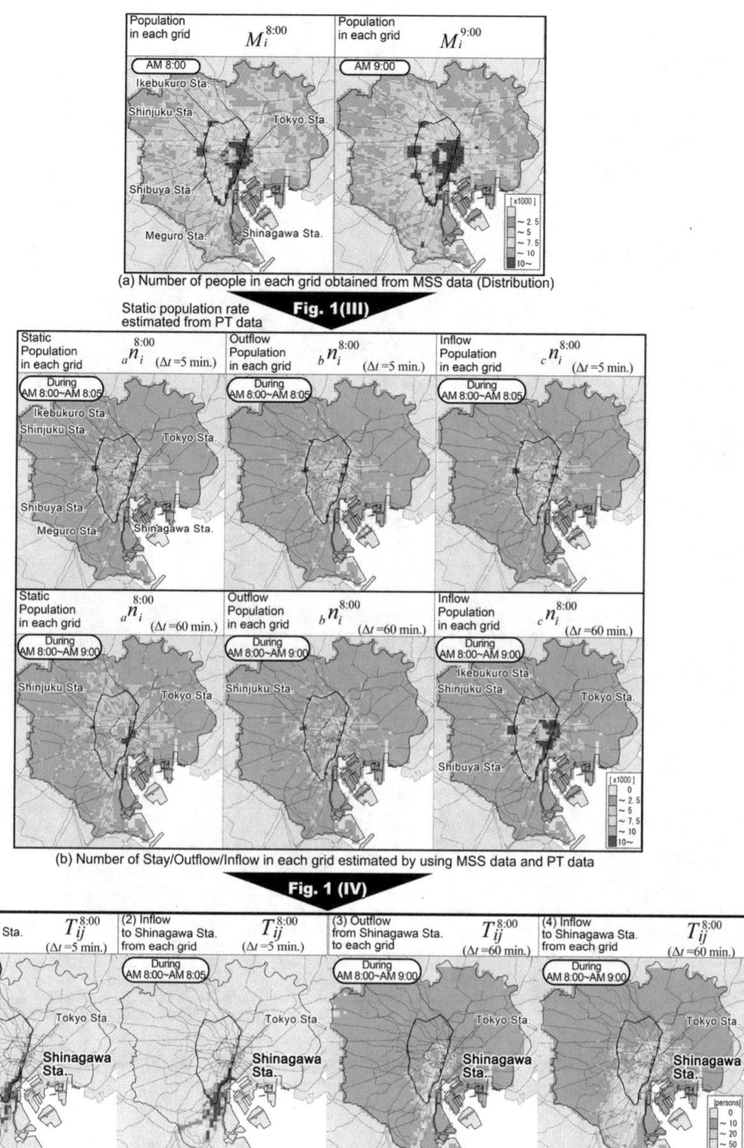

Fig. 6 Spatial distribution of Static/Outflow/Inflow population estimated using MSS data, Konzatsu-tokei, and PT data

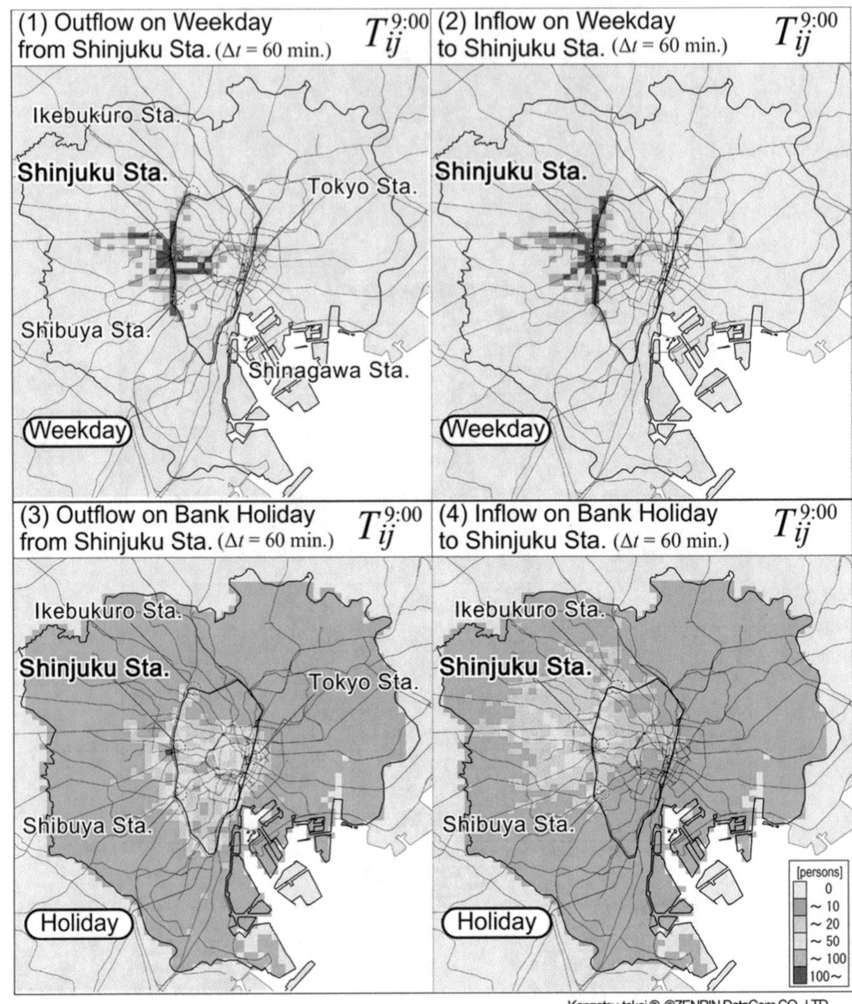

Fig. 7 Spatial distribution of the estimated moving population from/to Shinjuku Sta. between 9:00 and 10:00 on weekday and holiday

Due to human activities and mobility by rapid urban transportation systems, the distribution of population varies according to time. This is especially true in metropolitan areas [1].

Given these backgrounds, we discussed a method of integrating data on moving people (Konzatsu-tokei®) and regional population data (mobile spatial statistics), both of which are obtained from spatial distributions of mobile phone users and data on moving people based on questionnaire surveys (PT data). Using real data, we estimated numbers of moving people per 500 m grid-cell at any unit time and considered spatial movement such as moving directions and numbers of people flowing in/out of each grid-cell/at each time. The procedure proposed in this study makes it possible to identify the number of transient occupants and their travel directions at any time, on microspatial scales, by using the constructed spatiotemporal data for both static and transient urban occupants, and to obtain and use these basic data to analyze urban regions from new and never-before employed points of view.

The data used in this research are very rich, however, they are specific to the study area (Japan). Also, a discussion on the possible reusability including the challenges pertaining to it to other areas with poorer data would be very useful [22]. In further research, we would like to undertake a comparison of our proposed approach with relevant studies conducted in other countries addressing the same topic of people's movements. Also, MSS data used in this research is provided by one mobile phone operator. It would be interesting to elaborate on the implications of combining data from several mobile phone operators [23].

Using our proposed method, we would like to construct a model to evaluate the influence of large-scale public events or natural disaster on people's movements, which assists mitigating crowding and avoiding risks, identifying appropriate initial responses, and guiding evacuation [21]. The models proposed in the present study were created to offer basic data for a variety of analyses of urban areas. High-detail spatiotemporal distribution on human activities can be of great value in various fields, which include disaster risk management and simulation, regional and environmental planning, as well as geomarketing analysis.

Acknowledgements This paper is a part of the research outcomes funded by JSPS KAKENHI Grant Number JP17H00843. This report is an expanded and corrected version of Osaragi and Hayasaka (2019, 2020) [20–21].

References

1. Osaragi, T., Hoshino, T.: Predicting spatiotemporal distribution of transient occupants in urban areas. In: 15th AGILE Conference on Geographic Information Science, Lecture Notes in Geoinformation and Cartography, Bridging the Geographic Information Sciences, Springer, pp. 307–325 (2012)
2. Osaragi, T.: Estimating spatio-temporal distribution of railroad users and its application to disaster prevention planning. In: 12th AGILE Conference on Geographic Information Science, Lecture Notes in Geoinformation and Cartography, Advances in GIScience, Springer, pp. 233–250 (2009)
3. Osaragi, T.: Spatiotemporal distribution of automobile users: estimation method and applications to disaster mitigation planning. In: 12th International Conference on Information Systems for Crisis Response and Management (ISCRAM 2015). Proceedings of the ISCRAM 2015 Conference, ISCRAM 2015 Organization, May 2015 (2015)
4. Osaragi, T.: Estimation of transient occupants on weekdays and weekends for risk exposure analysis. In: 13th International Conference on Information Systems for Crisis Response and Management (ISCRAM 2016). Proceedings of the ISCRAM 2016 Conference, ISCRAM 2016 Organization, May 2016 (2016)

5. Sekimoto, Y., Shibasaki, R., Kanasugi, H., Usui, T., Shimazaki, Y.: PFlow: reconstructing people flow recycling large-scale social survey data. IEEE Pervasive Comput. **10**(4), 27–35 (2011)
6. Nakamura, T., Sekimoto, Y., Usui, T., Shibasaki, R.: Estimation of people flow in an urban area using particle filter. J. JSCE (D3) **69**(3), 227–236 (2013)
7. Hidaka, K., Ohno, H., Shiga, T.: Generating intra-urban human mobility and activity data by integrating multiple statistical data. J. JSCE (D3), **72**(4), 324–343 (2016)
8. Seike, T., Mimaki, H., Hara, Y., Odawara, T., et al.: Research on the applicability of "mobile spatial statistics" for enhanced urban planning. J. City Plan. Inst. Japan **46**(3), 451–456 (2011)
9. Seike, T., Mimaki, H., Morita, S.: Research on the evaluation of regional peculiarities in Kashiwa and Yokohama by "mobile spatial statistics". AIJ J. Technol. Design **21**(48), 821–826 (2015a)
10. Seike, T., Mimaki, H., Morita, S.: Study on the population characteristics in a city center district utilizing mobile spatial statistics: a new statistical method to capturing actual population distribution of the city through day and night. J. Architect. Plan. (Trans. AIJ) **80**(713), 1625–1633 (2015b)
11. Osaragi, T., Kudo, R.: Enhancing the use of population statistics derived from mobile phone users by considering building-use dependent purpose of stay. In: 22nd Conference on Geo-Information Science (AGILE 2019), Geospatial Technologies for Local and Regional Development, Springer, Cham, pp. 185–203 (2018)
12. Deville, P., Linard, C., Martin, S., Gilbert, M., et al.: Dynamic population mapping using mobile phone data. Proc. Natl. Acad. Sci. U.S.A. **111**(45), 15888–15893 (2014)
13. Ratti, C., Pulselli, R.M., Williams, S., Frenchman, D.: Mobile landscapes: using location data from cell-phones for urban analysis. Environ. Plan. B: Plan. Design **33**(5), 727–748 (2006)
14. Arimura, M., Kamada, A., Asada, T.: Estimation of visitor's number in mesh by building use by integrated micro geo data. J. Infrastruct. Plan. Rev. **33**, I_515-I_522 (2016)
15. Kamada, K.: Toshikotsubunnya ni okeru konzatutoukeideta no katsuyou ni tsuite, Meeting of Ministry of Land, Infrastructure, Transport and Tourism Kinki Regional Development Bureau **19** (2017)
16. Ishii, R., Shingai, H., Sekiya, H., Ikeda, D., et al.: A study about the improvement possibility of person-trip survey technique with mobile spatial dynamics. J. JSCE **55** (2017)
17. Matsubara, N.: Grasping dynamic population by "mobile spatial statistics": from the viewpoint of tourism disaster and stranded persons. J. Inform. Process. Manage. **60**(7), 493–501 (2017)
18. Calabrese, F., DiLorenzo, G., Liu, L., Ratti, C.: Estimating origin-destination flows using opportunistically collected mobile phone location data from one million users in Boston Metropolitan Area. IEEE Pervasive Comput. **10**(4), 36–44 (2011)
19. Iqbal, Md.S., Choudhury, C.F., Wang, P., Gonza'lez, M.C.: Development of origin-destination matrices using mobile phone call data: a simulation based approach. Transport. Res. Part C: Emerg. Technol. **40**, 63–74 (2014)
20. Osaragi, T., Hayasaka, R.: Estimating spatiotemporal distribution of moving people by integrating multiple population statistics. J. Architect. Plan. (Trans. AIJI) **84**(762), 1853–1862 (2019)
21. Osaragi, T., Hayasaka, R.: Prediction of spatiotemporal distributions of transient urban populations with statistics gathered by cell phones. In: Proceedings of the 6th International Conference on Geographical Information Systems Theory, Applications and Management—vol. 1: GISTAM, ISBN 978-989-758-425-1, pp. 33–44 (2020)
22. Kwan, M.P.: The uncertain geographic context problem. Ann. Assoc. Am. Geogr. **102**(5), 958–968 (2012)
23. Ricciato, F., Widhalm, P., Pantisano, F., Craglia, M.: Beyond the "single-operator, CDR-only" paradigm: an interoperable framework for mobile phone network data analyses and population density estimation. Pervasive Mob. Comput. **35**, 65–82 (2017)

A Digital Twin of the Social-Ecological System Urban Beekeeping

Carolin Johannsen, Diren Senger, and Thorsten Kluss

Abstract We describe the system design and setup of our digital twin of the social-ecological system urban beekeeping, with the aim to support agroecological methods in urban agriculture. The physical space consists of the bee populations, their beekeepers who are part of a beekeeping community, non-beekeepers who consume honey, organisational actors shaping rules and regulations and the environment. The virtual space is a multi-agent model, where autonomous agents can take actions and make decisions in partially observed Markov processes. To tie the physical and the virtual space, we embedded bee hives in an IoT environment and implemented an online documentation tool as a web application, where beekeepers take short notes about their work and observations. Bee hives are equipped with sensors, such as humidity, pressure and temperature sensors and a scale. Additionally, we pull data from the German weather service (Deutscher Wetter Dienst, DWD). In our system architecture, multiple levels on data fusion are performed, beginning with raw data quality estimation and sensor failure detection. On higher levels, states of entities are estimated, such as the health of a bee colony, and assessment made whether a state is normal or to be considered an anomaly. Finally on the highest level, we deal with the desires of our agents, how actions should be chosen in order to achieve or maintain desirable and rewarding world states. We hope to be able to refine our digital twin into a decision support tool for small-scale (bee) farmers and communal political actors that helps to reach desirable world states by predicting and simulating the effects of actions within the complex system of urban beekeeping.

Keywords Urban farming · Urban beekeeping · Agent-based modelling · Multi-agent models · Digital twin · Environmental modelling and simulation · Food supply system · Decision making · Decision support

---

C. Johannsen (✉) · D. Senger · T. Kluss
Cognitive Neuroinformatics Research Group, University of Bremen, Bremen, Germany
e-mail: C.Johannsen@uni-bremen.de
URL: http://www.cognitive-neuroinformatics.com/en

A. Kamilaris et al. (eds.), *Advances and New Trends in Environmental Informatics*, Progress in IS, https://doi.org/10.1007/978-3-030-61969-5_14

1 Introduction

We need to change the way we do agriculture. This is formulated in the Sustainable Development Goals set by the UN, in the synthesis report of the International Assessment of Agricultural Knowledge, Science and Technology for Development (IAASTD) initiated by the World Bank and co-sponsored by WHO, FAO and UNESCO [1], or in the recent 2019 FAO Report on "The State of the Worlds Biodiversity for Food and Agriculture" summarized in [2], to only name a few sources. There is a broad consensus that implementing biodiversity-friendly production methods and encouraging sustainable low impact practices by providing incentives for the responsible management of natural resources are key elements in this change process. Also, there is agreement on the benefit to focus on small scale farmers, with agroecological methods having been identified as one of the most robust [3]. As the trend for urbanisation is unbroken and the majority of the world's population will live in cities in the near future, it stands to reason to take a closer look at smart cities as the places where the suggested paradigm shifts and more sustainable practices could be implemented.

Mohammadi and Taylor suggest that digital twins can be a tool for (smart) cities to address the challenges of rapid urbanisation, as achieving urban sustainability and resilience objectives, including the planning and maintenance of water supply, recycling infrastructure, the allocation and consumption of resources [4]. Albeit they focus on human-infrastructure-technology interactions and do not explicitly include an edible green infrastructure [5] with non-human inhabitants and biodiversity in their perspective, the extension to include these aspects is straight forward. A city is a complex system of many and diverse actors, with a multitude of roles, tasks, functions, relationships and interactions. To capture this, the digital twin of a city can best be constructed as a composite, as an assembly of single twin entities as illustrated in [6]. This can be reflected in software architecture by a multi-agent system (MAS), which we will describe in more detail in Sect. 4.

Unlike unconscious digital twins in industrial contexts [7], our system includes digital twins of the inhabitants of the city, who are able to actively examine and deliberately alter their digital sibling through their behaviour. If used as a tool of empowerment for the citizens -small scale farmers and consumers alike—such a digital twin can be utilized for civic participation processes, not only reflecting the actions of each citizen in the citywide context, but visualizing them, giving the users an opportunity to virtually test the effects of their actions and behaviour changes. As Vickers and Grieves describe it in [8]: digital twins can help to decrease unpredicted undesirable outcomes by helping actors to better understand the system and their situation while the system is already in full operation and can not be stopped for testing.

Since the sustainability of urban agriculture largely depends on the institutional environment, including formal laws and regulations as well as social norms and rules [9], a planning and decision support system accessible to all stakeholders, urban farmers, political decision makers or institutions such as the veterinary office, seems profitable. The allocation and consumption of resources viable for sustainable urban agriculture such as land, water or subsidies [10] can be made transparent, which can be seen as an implementation of IAASTD's suggestion [1] to promote small-scale farmers by improving their access to political power, as transparency of political processes is improved.

A second dimension of the digital twin is leveraging sensor data of the agricultural processes and environmental conditions to improve the yield and resource usage, as well as observing the behaviour and health of plants and livestock, remotely and supported by AI [11].

2 Sensor-Supported Urban Beekeeping

In this paper, we focus on urban beekeeping as a subdivision of urban agriculture. Beekeeping has a long tradition in the area where we built our twin (Bremen, Germany) and is to date predominantly done by small scale farmers with less then 10 bee colonies under their care. None the less, a number of IoT systems for beehives exist [12–16] including our own development, described in more detail below and in [17].

A bee colony, with its division of work, different developing states of brood and well localised areas within the hive for different purposes, is in itself already a highly complex system. It can hardly be seen as an isolated entity, but as a component of an environment, dependent on climatic conditions, patch dynamics of the surrounding ecosystem, pollen availability, the density of bee populations in the area and current outbreaks of bee diseases. Its development strongly depends on the beekeepers, their experience and knowledge about bee biology and the decisions and beekeeping methods they perform.

Even though many beekeepers benefit from their experience of many years of beekeeping, decision making in beekeeping always remains a difficult task. The inside of a bee colony has a carefully managed mirco-climate, where the bees make sure that their precious brood can grow in optimal conditions and undisturbed by sunlight. Every time a beekeeper decides to make a manual inspection to asses the development and health of the colony, he/she disturbs this climate, a deregulation inside the habitation which the bees have to thoroughly readjust after the inspection. Adult bees as well as the brood are vulnerable to the exposure to cold temperatures, so that during winter or cold periods even a single opening of the hive might harm the colony. The use of sensors within the habitation -the hive- promises to be a possibility to observe changes of the bee development and health without major disturbances and the risk of cold exposure.

The beekeeping task is therefore comparable with complex systems occurring in the industrial domain where the concept of digital twins helped to establish a better monitoring environment, e.g. for intelligent workshop environments [18], product life cycle management [19] and aerospace vehicle testing [20].

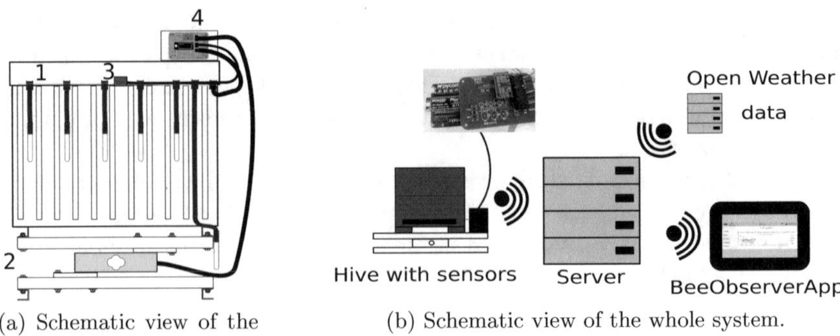

(a) Schematic view of the sensor hive.

(b) Schematic view of the whole system.

Fig. 1 Smart beehive sensor system

For the construction of the digital twin, we use a sensor monitoring system which was developed as Do-it-yourself (DIY) sensor kit as part of the citizen science project Bee Observer [21]. The advantage of a DIY sensor kit is that it is relatively cheap to purchase and can easily be transformed to match the needs rising from different bee hives and beekeeping techniques. As it was developed by citizen scientist with beekeeping background, it incorporates the experience of practical application of different earlier prototypes.

The sensor kit, schematically depicted in Fig. 1a, consists of six one wire digital temperature sensors (DS18B20), a single point load cell (Bosche H30A/ H40A) and a combined sensor for relative humidity, barometric pressure and ambient temperature (Bosch BME280). All sensors are connected to a circuit board and the measurements and the data transfer are performed by a ESP32 development board (FiPy, Pycom).

The scale (see Fig. 1a no. 2) is placed underneath the hive and the weight measurements can therefore reflect the nectar, pollen and water intake but also include the weight of the bees, the hive itself and snow or rainwater on the hive's roof. Five of the six temperature sensors (see Fig. 1a, no. 1) are placed within the hive, either all in the same clearance between two frames or one in every second clearance between two frames. Bees can regulate the temperature within the hive by quickly vibrating with the muscles of their wings to produce heat or by using their wings as ventilators to cool down the temperature. The optimal temperature to raise brood is 35 °C. Since the brood can only be found in a smaller ball-shaped area called the brood nest, it can happen that the temperature sensors are not optimally placed amidst the brood, so the uncertainty about the distance to the brood nest and the resulting temperature gradient must be accounted for. Other events, such as for example the advance of swarming, when half of the bee population emerges from the hive with a new queen and heat is generated when they prepare their muscles for the take off, are also reflected by changes in temperature measurements.

The combined relative humidity, temperature and atmospheric pressure sensor is placed within a clearance (Fig. 1a, no. 3). The bees constantly work on removing the water content from nectar to make it storable or on adding water to stored honey to

make it available to feed the brood, and there is an optimal humidity value in the hive for raising brood. Therefore, this value also is an important indicator for the state of the bee population.
As all climatic variables are influenced by the outside climate and weather, data from the open weather database of the German Meteorological Service is used as an additional source of information, fused with the measurements of one temperature sensor placed outside the hive and the atmospheric pressure sensor inside the hive.

3 Recording Bee Keeper's Actions and Expert Knowledge

To obtain direct information from the beekeepers, we built upon the online beekeeping application "BEEP", which was originally developed by a team in the Netherlands under an open source license [22].

The application is build to be a documentation tool for (urban) beekeepers. The beekeepers use the application to document objective information about their beekeeping activities, for example if they supplied additional food, harvested honey, or tried to treat a disease. They can also use it to communicate their own assessment of the population's health state, classified by items such as the activity of the bees, gentleness, pollen inflow, their brood status, abnormal observation such as bad wings, bad brood or diseases and parasites such as the varroa mite. The application provides the feature to file documentations for different locations of apiaries and different hive types. The users have the option to create groups for cooperative beekeeping, where group members share access to all hives of the group. The application also serves as a graphical user interface for the sensor measurements of the DIY sensor kit, which are visualised for each hive, which is shown in Fig. 2.

The entries made by the beekeepers about their actions and assessments are input for the digital twin, as they represent records of the behaviours of beekeepers and bee colonies.

4 The Digital Twin as a Multi-agent System

The domain we are modelling consists of multiple, diverse entities who are acting and interacting autonomously with each other and the environment. The model architecture has to be able to handle these properties. As formulated in the introduction, we see several use cases for the digital twin:

- Small scale beekeepers should be supported with the decisions in beekeeping. The decision support is based on inference on the bee colonies health and behaviour, on the grounds of sensor data as well as past observations and actions of the beekeeper.

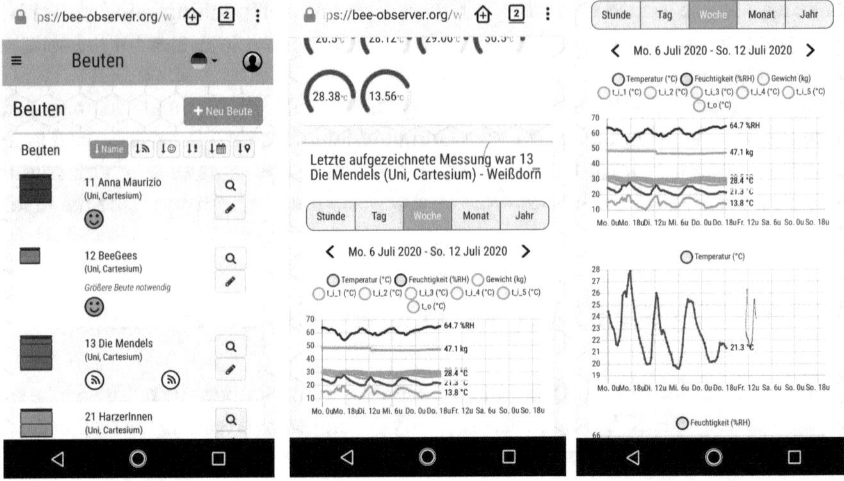

Fig. 2 The graphical user interface of the webapp which is used for documentation and to view the sensor readings

- Political decision makers, such as members of the city council or officers in the veterinarian agency, should be able to simulate the effects of their actions, possible changes in rules and regulations, with the goal of avoiding undesirable system behaviour.
- Citizens should be able to see the allocation of resources such as funds or the rights granted to keep bees in certain locations.

Therefore, we defined the requirement specification that the system architecture is modular, so that a single beehive and each small scale farmer is individually represented, but at the same time the dynamic macro-scale behaviour of an entire city can be observed and simulated. A modelling approach which fulfils this requirement is agent based modelling. It comes with the bonus of being relatively intuitive for users, as the entities represented as individual, autonomous agents often have a limited repertoire of actions, from which the macro-scale system behaviour emerges. Multi-agent systems have been successfully applied in a wide range of application contexts [23], including sociology [24], socio-technical systems modelling [25] and human-nature or social-ecological systems [26, 27].

To be able to run simulations and explore system behaviour in regions of the state space that have previously not been reached, the digital twin does not only need a representation of the sensor data and system states, as for example shown in [11], but also an AI component for the decision making process of each agent, including the communication between agents.

In our architecture, each agent holds a probabilistic representation of the world and its state, called the agents' beliefs. The agent constructs its internal representation of the world out of information it perceives coming from the environment or from other agents. This information is not complete, each agent only possesses a limited number of senses and resources to perceive the world, or in other words, the world is only partially observable. We further make the Marcovian assumption, that an agent believes the current state of the world to be (solely) dependent on the state in the time step immediately before. The states of the world are denoted as a finite set S, with $s_t \in S$ as the the state at time t. All observations of those states are denoted by O. An agent does not blindly believe all information to be true, but does believe that its view of the world is true, which it updates after making an observation. The agent has a mental concept of how likely it is that a certain observation occurs under the given circumstances, its observation model, defining the probability of measuring $o \in O$ given the world is in state $s \in S$. It also has a mental model of how the world changes from one time step to the next, which is often referred to as the transition model, defining the probability of reaching the state $s_{t+1} \in S$ given a starting state $s \in S$ and performing an action $a \in A$. The transitions depend on the actions A of the agent. Both of these mental concepts are individually learned and dynamically change through life experiences. However, in the current version of our architecture they are time-invariant. In short: we designed a partially observed Markov decision process (POMDP) model as a dynamic Bayes net as compact representation of the beliefs. See, for example [28] for a more detailed explanation.

At each time t, an agent has the opportunity to take some actions that will have an impact on the state of the world. Based on expert interviews, literature review and the data collected through the BeeObserverApp, we have constructed a library of actions A which each agent can chose from. These actions can be sequentially combined into a policy π. Even though not all actions do require the same amount of time, they are atomic in a sense that an agent will not abort an action once it has started it. Even if some actions have a duration longer than one time step, the agent will finish the action and only afterwards update its beliefs and revise its current policy if necessary. The action library depends on the type of agent: *bee colony*, *beekeeper*, *non-beekeeper* (both of which are individual humans) and *organisation*. It is describes in more detail in [29], but can be summarized as follows:

Bee colonies

consume: Eat from the honey/syrup stored in the cells; **forage:** If the weather permits, the foragers of the colony will fly out into the neighbourhood and gather nectar and thereby replenish the food stock; **storeSyrup:** If a beekeeper has provided syrup, this is taken from the container and stored within the hive as food stock; **raise queen:** If a colony is in need of a new queen, a queen cell is made; **swarm:** Half the colony leaves with the old queen to find a new home. The other half stays and keeps on living in the same hive; **die:** A colony can die;

Beekeepers

inspect: In order to get information about the status of the colony, their health, food supplies and so on, the beekeeper performs an inspection; **feed:** Providing sugar syrup as a food supplement; **make new colony:** An equal fraction of all of the hives in care of the beekeeper is taken and combined to form a new colony; **register:** Each colony must be registered with the veterinary office once it comes into the possession of a beekeeper; **unregister:** A colony must be unregistered with the veterinary office once a beekeeper no longer owns it (also in the case of death); **break queen cell:** If a queen cell is observed, the beekeeper can chose to destroy it (and frequently does so, to prevent swarming, for example); **combine colonies:** Two colonies are merged into one; **harvest:** Take honey out of the hive; **treat:** The parasite varroa destructor is a lethal threat to colonies. Treatment with medication such as organic acids, improves colony health; **sell a colony:** If a buyer for the colony can be found in the same or directly adjacent neighbourhoods; **buy supplies:** Medication, sugar syrup or empty hives, provided sufficient monetary funds; **sell honey:** We model a beekeeper to always consume what he/she needs for her own subsistence first and sell excess only; **remove hive:** If a colony dies, it's remains need to be taken care of. Possibly, the honey it has left behind could be harvested. The now empty hive is stored for reuse;

Beekeepers and non-beekeepers
catch the swarm: If a colony swarms, the swarm flies away and any human who sees it can catch the swarm. This will turn the human into a beekeeper, if he/she has not been one; **eat:** An individual amount of honey every simulation time step; **buy a colony:** From a beekeeper in the same or immediately adjacent neighbourhood. This will turn the human into a beekeeper; **attend a beekeeping course:** To increase his/her knowledge on beekeeping. A course can be offered by an organization; **meet:** A random number m [0..50] of humans from its neighbourhood, sequentially. If a consumer and a beekeeper meet, a honey transaction occurs if the consumer is willing to buy and the beekeeper has honey in stock.

Organisations
There are currently only a limited number of organisations implemented. Each organisation has a very small and highly individual action library.
The *veterinary office* can **prohibit** the placement of a new colony in the neighbourhood. This is a reaction to an registration attempt. It can also **set colony density**, setting a maximum number of colonies allowed in a neighbourhood. The *beekeepers association* can **offer a beekeeping course**.

The decisions of the real world twins are captured through sensor data, yet one application of the digital twin is to predict and simulate how changes will affect the systems' states in the future. To be able to make such predictions, a decision making process needs to be implemented. Since we are dealing with a complex, dynamic environment, our decision making architecture should be able to predict dynamic, non-linear behaviour. A particle filter can be used to infer what the agent believes to be the future states of the world, under a chosen policy π. Out of all possible states of the present and the future, only a small subset is considered by the agent. For these states, particles are generated [30–33].

To be able to infer which decisions an agent should take, a reward (or utility) function R is defined. In many works on POMDPs, the rewards are assigned to each action. But for a finite-horizon problem, this is equivalent to assigning a reward to states [34]. As we believe a majority of people would consider it desirable to work less and not more, we chose to reward states instead of actions. The agent plans its policies by maximising the total reward over a time horizon τ (or to infinity)

$\sum_t^{t+\tau} R(s_t, a, s_{t+1})$ [35]. It is a common practice to calculate this iteratively, so that for each time step the maximum value is chosen, starting with the very last action of the sequence at time $t = \tau$. There is very little research published on how humans or non-human animals determine a feasible length of a policy.

In our setting, as in many real world settings, there might be multiple objectives (or desires) an agents holds. For example, a beekeeper might want to keep his/her bees healthy, but also make a profit by selling honey. While it is often possible to restructure the reward function from a multi-objective into a into a single-objective case [36], the question remains how to set the rewards, how to determine what are desired states. Following a homeostasis approach [37], we have defined the initial goal states of *staying alive* and *procreating* for the bee colonies, *keeping the colonies alive* as goal for beekeepers as well as *honey self sufficiency* and *maximising profit* for beekeepers and non-beekeepers alike. These goals need to be refined as more data becomes available and to be extended to the motivations of organisational actors.

5 Connecting the Twin to Sensors

5.1 Data Storage and System Architecture

The measurement readings of the sensors in a single bee hive are initiated by the ESP32 micro-controller every 5–10 s. Each sensor-kit is identified by a unique key. The vector of sensor measurements, consisting of seven temperature values, one humidity, one barometric pressure and one weight value, is transferred to an influxDB server using an HTTP API, associated with the unique sensor key and the current timestamp. The influx database is a no-sql database optimised for time-series with a high writing and querying throughput rate [38]. It is open source, implemented in Go and has a sql-like query language, featuring special filters and aggregations for the time-series data [39]. From the MAS, which is implemented in java and R, sensor measurements can be directly read using the influxdb http-api.

All information entered in the online BeeObserver application is stored in a MySQL database. For every hive equipped with a sensor-kit, a separate virtual hive is created in the app, identified by the unique sensor-kit key. A hive can be associated with an apiary, a group of hives at the same physical location. The user who created the virtual hive (the owner) can invite other beekeepers in the app for cooperative beekeeping, and all of these beekeepers are allowed to enter inspections. Therefore, there are different databases for different entities: hives, apiaries and inspections. The hive database includes information such as type and size of a hive and details about its bee population, such as the age of the queen bee. The apiary database can be accessed for the detailed location of an hive. The inspection database can be accessed regularly to get updates about performed actions of the bee keeper and their assessment of the bee populations health and development state. Current and forecast weather data is obtained from the German weather service (Deutscher Wetter Dienst, DWD) [40].

The integration of online Machine Learning and classical time series analysis are implemented in the statistical programming language R, while the simulation of the multi-agent system model with graphical user interface is implemented in java.

For illustration purposes, we show measurements for two exemplary events, a swarming event and the feeding by a beekeeper. During a swarming event in a hive with a new queen, half of the colony leaves the hive with the old queen, leading to a sudden decrease in weight. Also, the temperature continuously increases when bees prepare and gather to leave the hive, having its peak when most of the bees left already (Fig. 3).

To feed a bee colony, the beekeeper has to open the hive, leading to a decrease in weight. Next, an additional super is added together with syrup or fondant. After the feeding, the weight should decrease as the bees extract the sugar from the fed supplement and evaporate the water content (Fig. 4).

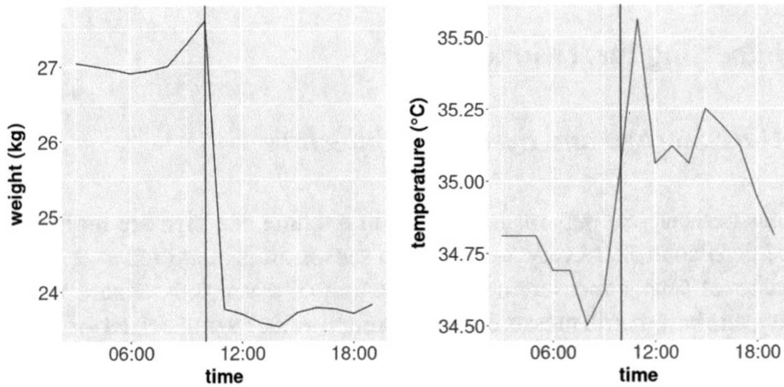

Fig. 3 Sensory data from bee hive in Bremen, Germany, showing swarming event, with queen and half of the colony leaving the hive. Vertical line: Onset of weight decreases/ estimated starting time of swarm. Left: Total weight in kg. Right: measurements of central temperature sensor (degree Celsius)

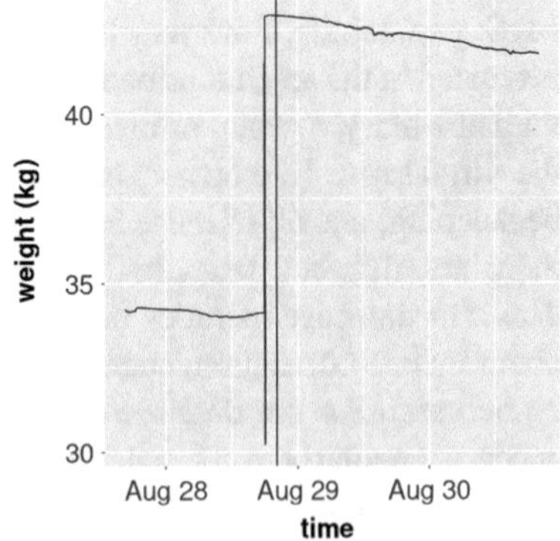

Fig. 4 Weight measurements before, during and after an inspection with feeding of a bee colony in Cologne, Germany. Vertical line indicates the reported point in time of the inspection by the beekeeper

Referring to the levels of data fusion [41, 42], we have to perform multiple levels of data fusion. On level 0, we need to estimate the quality of our raw sensor readings, filter missing data, check for faulty or failed sensors and perform temporal data alignment. This includes inference about the time delay between an event observed by a beekeeper and the documentation thereof. Often, the time precision of a beekeeping intervention record is within the fraction of an hour, and not seconds, as digital sensor readings with timestamps (Fig. 4). Also, an observation of a longer lasting state, such as a queen bee missing, might occur some time after the onset of the event. On level 1, we estimate entity states and features, such as the health of a colony or the amount of food resources the colony has left, if a queen is present or if a beekeeper is currently issuing a treatment against an infection. On level 2, we perform the situation assessment, where we estimate whether the current state constitute an immediate or future problem. In this step, we regard the relationships between entities and their environment. On level 3, we deal with the desires of our agents, how action policies π should be chosen in order to achieve or maintain desirable and rewarding world states.

6 Decision Making Example

In this section, we describe one decision process relevant in urban beekeeping in more detail. After a harvest or when there is a shortage in forage, beekeepers feed their colonies with sugar syrup. This is a source for optimising profitability of a beekeeping operation [43] and an issue which came up during our expert interviews. The colony should only be fed the exact amount it needs, so no money is invested in syrup not needed by the hive, no time is wasted by needlessly refilling containers and no predators are attracted. But the colony also should be provided with the amount it needs, as a failure in feeding might result in suffering and long term negative health effects due to food deprivation. Arguments against a constant provision with sugar syrup are: (1) Syrup might turn bad if it is not brought into the hive in time. Even if it is still good, it can not be put back into storage when it is not collected by the bees. (2) Syrup might be left over after the winter, when the colony is already foraging. The syrup must be removed from the cells before harvesting, else it contaminates the honey. (3) Syrup fed in summer can attract robbery by parasites, such as wasps.

The knowledge on the proper amount to feed is implicit to many experienced beekeepers, but has been identified as an issue where decision support would be valuable. Two decision scenarios have been modelled: (1) How much syrup should be fed in autumn as preparation for overwintering. (2) When and in what quantity to provide syrup during spring or summer as an intervention to prevent malnutrition.

Figure 5 shows the Bayes net for the inference process. The decision support will run a query on the probability of gSyrupNeeded, a continuous node modelling the needed quantity of syrup in gram. This depends on the provisions the colony has left, the population size, and the time of year. Neither the provisions nor the population size can be measured directly, but indirectly. The population size is subject to the

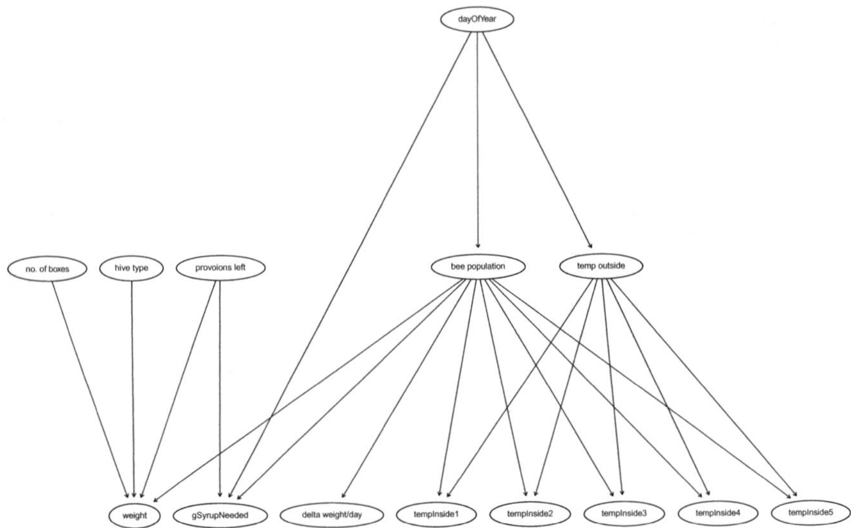

Fig. 5 Bayesian network to infer how much syrup should be fed

biological rhythm of the bees and therefore depends on the time of year. But of course there are weaker and stronger colonies, which can be indirectly measured through the amount of food they consume per day and the size of their brood nest. To approximate the food consumption, we introduced an additional, virtual sensor (denoted as *delta weight/day*in the graph). We calculate this by comparing the weight readings at midnight every day (inducing only a minor, time constraint dependency to the scale measurements). The brood nest size affects the temperature within the hive, which can be captured through the temperature sensor array. Note that we have omitted here the level 0 data fusion actions: the input of the temperature sensors and the scale is already checked and corrected for sensor failures and poor quality data, the timing of observations has been aligned with the sensor-kit readings and the calendar is accessed correctly.

How many provisions are left can also only be measured indirectly, as a fraction of the total weight of the hive. The weight constitutes of the empty housing, which depends on the *hive type*. This is documented in the BeeobserverApp. Since some hives are of the Langstroth type, made out of modular stackable boxes, the number of boxes is variable and changes during the different phases of the beekeeping year. The alteration of the hive (stacking, taking off of boxes) should be documented in the app. Precipitation is also an influencing factor, since all bees stay inside the hive when it is raining, and there could be - at rare times in Bremen- a layer of snow on the hive's roof.

The net represents one distinct timepoint t. All dynamics of the state transition from t to $t+1$, given an action a_t, is modelled in the transition model T, with probabilities $P_{as}(s_{t+1})$. For example, how much food will be gathered by the bees is such an action a, affecting a subset of world states, including *provisions left*, but the amount depends on the value of several state variables, such as the outside temperature, precipitation, day of year, bee population and (not modelled here) the number of colonies in the vicinity and available forage.

The parameters of the probability distributions have been learned from the data we have already collected and supplemented with expert knowledge. However, these parametrizations can be regarded as preliminary only, as the data collection within the hives is subject to the yearly rhythm of the bees and surrounding nature.

7 Conclusion and Future Work

We have set up the digital twin, connecting the physical world of bee colonies and beekeepers to a multi-agent system and designed the system architecture which can handle the different levels of data fusion. The machine learning layer is capable of constantly learning and updating the conditional probabilities of the states and state transitions but needs much more data to meaningfully do so. The effects of different weather and climate conditions in combination with diverse beekeeping choices, for example, should be recorded for a number of individual colonies before a generalization can be made. This data will be collected with the sensors already in place, and the BeeObserver project is in the process of bringing more sensor-kits to the field, slightly slowed by Covid-19, but the time frames for data collection are subject to the rhythm of nature.

Many details of the implementation remain for future work, such as the modelling of all relevant decision processes and belief updates for all agent types, as well as identifying, predicting and assessing many more entity states, such as preparation for swarming or loss of a queen.

There is currently no data transfer from the real world to the agent-based model regarding the human interactions. Validation of the modelled system dynamics and simulation results will require additional data acquisition. We are currently investigating the feasibly of an interactive gamified approach, where the simulation of urban beekeeping can be played just like a farming simulator.

We wish to thank all citizen scientist of the Beeobserver project for their hard work and positive energy invested into the set-up and maintenance of the sensor network.

References

1. S. International Assessment of Agricultural Knowledge, T. for Development, Agriculture at a Crossroads - Synthesis Report. Tech. Rep. (2009)
2. Pilling, D., Bélanger, J., Hoffmann, I.: Nature food **1**(3), 144 (2020). https://doi.org/10.1038/s43016-020-0040-y

3. Altieri, M.A., Funes-Monzote, F.R., Petersen, P.: Agronomy for sustainable development **32**(1), 1 (2012). https://doi.org/10.1007/s13593-011-0065-6
4. Mohammadi, N., Taylor, J.E.: In: 2017 IEEE Symposium Series on Computational Intelligence, SSCI 2017 - Proceedings **2018-Januay**, 1 (2018). https://doi.org/10.1109/SSCI.2017.8285439
5. Russo, A., Cirella, G.T.: Palgrave Commun. **5**(1), 1 (2019). https://doi.org/10.1057/s41599-019-0377-8
6. van Schalkwyk, P., Malakuti, D.S., Lin, S.W.: IIC J. Innov. (November) **2** (2003)
7. Tao, F., Qi, Q.: Nature **573**(7775), 490 (2019). https://doi.org/10.1038/d41586-019-02849-1
8. in *Transdisciplinary Perspectives on Complex Systems: New Findings and Approaches*, August 2017 (2016), pp. 1–327. https://doi.org/10.1007/978-3-319-38756-7
9. Pearson, L.J., Pearson, L., Pearson, C.J.: Urban agriculture: diverse activities and benefits for city society **5903**, 7 (2011). https://doi.org/10.3763/ijas.2009.0468
10. Cohen, N., Reynolds, K.: Renewable Agric. Food Syst. **30**(1), 103 (2015). https://doi.org/10.1017/S1742170514000210
11. Alves, R.G., Souza, G., Maia, R.F., Tran, A.L.H., Kamienski, C., Soininen, J.P., Aquino, P.T., Lima, F.: In: 2019 IEEE Global Humanitarian Technology Conference, GHTC 2019 (October) (2019). https://doi.org/10.1109/GHTC46095.2019.9033075
12. Edwards-Murphy, F., Magno, M., Whelan, P.M., O'Halloran, J., Popovici, E.M.: Comput. Electron. Agric. **124**, 211 (2016). https://doi.org/10.1016/j.compag.2016.04.008
13. Pešović, U., Marković, D., urašević, S., Ranić, S.: Acta agriculturae Serbica **24**(48), 157 (2019). https://doi.org/10.5937/aaser1948157p
14. Chen, Y.L., Chien, H.Y., Hsu, T.H., Jing, Y.J., Lin, C.Y.: Y.C. Lin. In: Yang, C.N., Peng, S.L., Jain, L.C. (eds.) Security with Intelligent Computing and Big-data Services, pp. 535–543. Springer International Publishing, Cham (2020)
15. Catania, P., Vallone, M.: Sensors (2020). https://doi.org/10.3390/s20072012
16. Hunter, G., Howard, D., Gauvreau, S., Duran, O., Busquets, R.: Proc. Inst. Acoust. **41**(June), 339 (2019)
17. Johannsen, C., Senger, D., Kluß, T.: In: 2020 16th International Conference on Intelligent Environments (IE) (2020)
18. Zhang, Q., Zhang, X., Xu, W., Liu, A., Zhou, Z., Pham, D.T.: In: International Conference on Intelligent Robotics and Applications, pp. 3–14. Springer, Berlin (2017)
19. Tao, F., Cheng, J., Qi, Q., Zhang, M., Zhang, H., Sui, F.: Int. J. Adv. Manuf. Technol **94**(9–12), 3563 (2018)
20. Glaessgen, E., Stargel, D.: In: 53rd AIAA/ASME/ASCE/AHS/ASC structures, structural dynamics and materials conference 20th AIAA/ASME/AHS adaptive structures conference 14th AIAA (2012), p. 1818
21. Bee Observer BOB das ist unser citizen science projekt. https://hiverize.org/bee-observer-bob-das-ist-unser-citizen-science-projekt/. Accessed 01 July 2020
22. (2020). https://beep.nl/home-english. Accessed on 15 Jan 2020
23. Dorri, A., Kanhere, S.S., Jurdak, R.: IEEE Access **6**(April), 28573 (2018). https://doi.org/10.1109/ACCESS.2018.2831228
24. Bianchi, F., Squazzoni, F.: WIREs Comput Stat **7**(August) (2015). https://doi.org/10.1002/wics.1356
25. Vespignani, A.: Nat. Phys. **8**(1), 32 (2012). https://doi.org/10.1038/nphys2160
26. Schulze, J., Müller, B., Groeneveld, J., Grimm, V.: J. Artif. Societies Soc. Simul. **20**(2), 8 (2017). https://doi.org/10.18564/jasss.3423. http://jasss.soc.surrey.ac.uk/20/2/8.html
27. An, L.: Ecol. Model. **229**, 25 (2012). https://doi.org/10.1016/j.ecolmodel.2011.07.010
28. Thrun, S., Burgard, W., Fox, D.: Probabilistic Robotics. Massachusetts Institute of Technology (2006)
29. Johannsen, C.: In: Under Consideration for 24th European Conference on Artificial Intelligence, Qualitative Reasoning Workshop. Springer, Berlin (2020)
30. Doucet, A., Johansen, A.M.: Handbook of Nonlinear Filtering (December) **4** (2009)
31. Doshi, P., Gmytrasiewicz, P.J.: In: Proceedings of the International Conference on Autonomous Agents, pp. 463–470 (2005). https://doi.org/10.1145/1082473.1082522

32. Bard, N., Bowling, M.: Proceedings of the National Conference on Artificial Intelligence **1**, 515 (2007)
33. Arulampalam, M.S., Maskell, S., Gordon, N., Clapp, T.: Bayesian bounds for parameter estimation and nonlinear filtering/tracking **50**(2), 723 (2007). https://doi.org/10.1109/9780470544198.ch73
34. Kaelbling, L.P., Littman, M.L., Cassandra, A.R.: Artif. Intell. **101**, 99–134 (1998). https://doi.org/10.1007/s00726-010-0654-8
35. Ng, A., Harada, D., Russell, S.: ICML **99**, 278 (1999)
36. Roijers, D.M., Vamplew, P., Whiteson, S., Dazeley, R.: J. Artif. Intell. Res. **48**, 67 (2013). https://doi.org/10.1613/jair.3987
37. Drengstig, T., Jolma, I.W., Ni, X.Y., Thorsen, K., Xu, X.M., Ruoff, P.: Biophys. J. **103**(9), 2000 (2012). https://doi.org/10.1016/j.bpj.2012.09.033
38. (2020). https://docs.influxdata.com/influxdb/v1.7/. Accessed on 29 Apr 2020
39. Nasar, M., Kausar, M.A.: Int. J. Innov. Technol. Explor. Eng. **8**(10), 1850 (2019)
40. O.D.S. DWD, *Data Source: Deutscher Wetterdienst* (2020). http://shorturl.at/lpsV2. Accessed 28 Jan 2020
41. Llinas, J., Bowman, C., Rogova, G., Steinberg, A., Waltz, E., White, F.: In: Proceedings of the Seventh International Conference on Information Fusion, FUSION 2004 **2**, 1218 (2004)
42. Steinberg, A.N., Bowman, C.L: pp. 1–18 (2004). http://www.infofusion.buffalo.edu/tm/Dr.Llinas'stuff/RethinkingJDLDataFusionLevels_BowmanSteinberg.pdf
43. Somerville, D., Collins, D.: **63**, 2007 (2015)

Utilizing CityGML for AR-Labeling and Occlusion in Urban Spaces

Patrick Postert, Markus Berger, and Ralf Bill

Abstract Cities are one of the most pertinent application spaces for current Augmented Reality systems but have proven to be technically challenging. They are abundant with information, while also requiring occlusion of that same information at different levels of depth. With more and more detailed 3D city models becoming available through the CityGML standard, we introduce an approach that can display contextual multimedia annotations on buildings in AR and properly occlude them without depending on complex computer vision analysis. Our approach also integrates directly with CityGML through an Application Domain Extension (ADE). We implement a prototype that allows for the dynamic textual labeling of urban environments on a variety of AR-capable smartphones using the Unity game engine.

Keywords Augmented reality · CityGML · ADE · Annotations · Label placement · Occlusion · Unity · ARCore · ARKit

1 Introduction

1.1 Motivation

In current Augmented Reality (AR) technology, there are several underlying problems that every system and application has to consider. What information is important enough to display in the limited screen space? Which parts of the real world should virtual objects be allowed to occlude, and can real objects hide virtual ones as well?

P. Postert · M. Berger (✉) · R. Bill
University of Rostock, Rostock, Germany
e-mail: markus.berger@uni-rostock.de

P. Postert
e-mail: patrick.postert@uni-rostock.de

R. Bill
e-mail: ralf.bill@uni-rostock.de

A. Kamilaris et al. (eds.), *Advances and New Trends in Environmental Informatics*, Progress in IS, https://doi.org/10.1007/978-3-030-61969-5_15

How precisely does a user need to be located? All these questions also heavily depend on available hardware and software capabilities.

For this paper, we specifically try to look at the use case of showing contextual information about an urban environment, while the user is located within it. Labels and graphics on buildings as well as routes on the street and walkways may be the most prominent examples, but there are many possibilities. What almost all of them have in common is that they are related to some object in the environment, usually a building or a street. The more precise our idea of these objects is, the better the digital content can be positioned relative to them. One particularly important part of alignment in these cases is proper occlusion. Near-field, real-time occlusions have become available even on mobile devices through real-time depth estimation [13]. However, the far-field is a lot more challenging and requires computationally complex computer vision algorithms like semantic image segmentation [15], which are often prone to errors and inaccuracies. Without occlusion, annotations become unusable in a dense city. Users may be able to see labels for buildings on the other side of a city block, which visually clutter the annotations on their side. To solve this, the clipping distance for virtual objects would have to be drastically limited, which could result in relevant annotations that should be visible but are further down a street to vanish. The problem only compounds when annotations need to be placed at different heights. If we want to label a faraway skyscraper, how do we know when users can see it over the roofs of other buildings? The only reliable way to do this is to use precisely located 3D building geometry.

Unfortunately, most commonly available city data sets only feature buildings as 2D polygons with a single height value. At the same time, individual objects (like lamp posts or bike stations) only have one pair of coordinates without any shape information. If there is any detailed 3D geometry, it is rare and usually reserved for specific areas or landmark structures. The most prominent and easily accessible data set of this kind is the OpenStreetMap (OSM) database. Moreover, companies like Google also offer APIs for regulated and paid access to their spatial data.

The most significant capacity to offer detailed 3D data for a specific city or region still lies with local administrations. If they wish to do so, they can combine data from several different sources and arrange detailed surveying where necessary. One way to store and publish this information is through the CityGML standard. It offers several levels of detail, ranging from simple 2D polygons to fully-featured building models with interior. The latter end of the spectrum is not widely available, but data sets with detailed 3D exteriors are becoming more popular.

In this paper, we will present a way to utilize CityGML to show properly occluded, contextually positioned annotations in urban environments. We will focus on the technical aspects of this problem, and mainly consider how to add labels to buildings. For this purpose, we define labels as a type of annotation consisting only of text that does not allow for user interactions. However, we make note about how to extend the system where necessary.

1.2 State of the Art

The idea to utilize CityGML, or city models more generally, in AR applications is not new. Blut et al. [3] describe how to store, process, and serve the potentially large data sets on a mobile android device. However, so far these models have most frequently been used for visualization of the building information itself, or for outdoor positional tracking, either by matching real and virtual building edges [12] or by segmenting a camera image and fitting it to the model [7]. Recently, Blut and Blankenbach [2] used CityGML data for global pose tracking in both indoor and outdoor environments.

Therefore, for our system, we assume that precise tracking could be implemented according to one of these methods for generalized, real-world use. For the prototype, we emulated a similar quality by placing image targets into the environment.

The better these tracking systems become, and the more readily available AR is on mobile devices, the more pressing the need for new information display techniques gets. Several issues are presenting itself: How to render text at changing distances, how to make information visible in bright sunlight, how to deal with shifting backgrounds, and how to keep the user aware of their surroundings. There are also questions of authoring, registering content to the environment, and user interfaces [8].

The issue of label-placement is generally more well-researched than the other problems, both for immersive display systems [1] and even longer for 2D layout systems and cartography [5]. Every 2D labeling system must deal with overlapping labels and every 3D system with self-occlusions. For example, McNamara et al. [11] investigated how to label 3D objects that are located very close to each other. The study ascertained that presenting only labels associated with objects the user is focusing on leads to a faster search for information while also considerably reducing the risk of overlap, showing that a reduction in content is often preferable. The authors used a virtual reality eye-tracking enabled head-mounted display for their evaluation, but the principal conclusions are transferable to AR applications, as cross-label occlusion behaves the same.

2 Materials and Methods

2.1 Requirements

The proper choice of data format is essential for the integration of building models into AR environments, as broad availability of real-world information is crucial for the comprehensive labeling of urban spaces. A widespread format that is also utilized by authorities and government agencies in Germany is CityGML [10]. CityGML is an XML-Schema based on the Geographic Markup Language (GML) and is used as a storage and exchange format for 3D city geometries [3]. Its most common use case is collections of buildings, but it can also handle streets and other types of objects. CityGML geometries are boundary representations. Through a Level of Detail (LoD) concept, these objects can be stored at multiple levels of fidelity. LoD 0 features 2D building shapes with one associated height value, LoD 1 consists exclusively of geometries extruded upwards from 2D, LoD 2 stands for detailed wall and roof shapes, and LoD 3 extends this to structures like windows and balconies. CityGML, however, is an evolving standard, and there are moves towards a more flexible naming scheme [9].

Barring any especially faulty or low-quality data sets, CityGML building models are of a high enough fidelity to allow for accurate occlusions between buildings and annotations. To correctly position buildings and annotations, we also need to place and precisely locate the image targets both in the real world and in the data set. For this reason, the utilized coordinate system either needs to be the same one used in the respective CityGML data or we need to reproject our data beforehand.

2.2 Hardware and Software

Over the last few years, a significant part of all AR applications has been developed with game engines. Despite the extensive tools they offer, they are often freely available for non-commercial applications. We used Unity for this study as it allows a cross-platform deployment, an object-orientated approach for modeling and implementation, as well as the possibility to integrate compiled *Managed Plug-ins* or *Native Plug-ins*. The former are limited to the .NET libraries and the C# language within the Unity framework, the latter benefit from a deep integration into the operating system of the target devices and can embed software components written in other programming languages such as C, C++, Java, or Objective-C. If the plug-in supports the desired target platforms, this is not restricting cross-platform development.

A variety of extension exists to add AR functionality to Unity. With *AR Foundation*, Unity Technologies supplies a free plug-in, which utilizes the OS-level AR capabilities available on recent smartphones in the form of Google's *ARCore* and Apple's *ARKit*. This grants access to most of the mid-range and high-end smartphones and tablets released in the last years, as most iOS devices presented since 2015 support *ARKit* and a substantial amount of Android devices support *ARCore*. We implemented and evaluated our prototype using an Apple *iPhone 7 Plus*, an HMD Global *Nokia 7 Plus*, and a Samsung *Galaxy S8*.

With data sets as large as whole cities, storage becomes an especially important question on mobile devices. Local storage space is scarce, but stable network connections are also not ensured in most areas of the world. Ideally, users would download the parts of a data set they are likely to encounter while they are in a stable network and store it on the device until needed. How this selection is done heavily depends on the specific application domain of an AR system. For everyday use, it could be as simple as downloading everything in a radius around the user location every time there is a Wi-Fi connection, thus covering both common routes near their home or workplace, as well as new locations they arrive at. For many professional applications, the area of interest will be known before the expert user gets there, thus the offline data set could be refreshed each day depending on upcoming tasks.

Whatever approach users would ultimately take, the data must be both stored and later loaded as efficiently as possible, and any loading strategy needs to work on all target devices. To achieve this, we use the *SQLite* database management system (DBMS), which has little overhead and implements most of the common database operations needed here. *SQLite* can be integrated into a Unity project through the game engine's *Native Plug-in* interface. At the time of writing, there were no available Unity implementations for *SQLite*'s spatial extension *SpatiaLite*, which would simplify some of the caching and querying operations.

If stable network speeds can be ensured, it would also be possible to serve CityGML data over standardized interfaces like Web Feature Services (WFS), which also operate on a GML encoding scheme. The necessary spatial operations could be represented with filtered requests, or if more complex through a specialized Web Processing Service (WPS) operation.

2.3 Label Placement

There is a wide range of semantic information in CityGML models. Depending on the use case, some of this information could already be used to generate labels automatically. However, as mentioned before, automatic labeling is a complex problem, and because AR always alters the user's perception of reality, we should strive to be as deliberate as possible. Simply rendering a street name and building number on every street-facing surface would have little benefit and would result in needless clutter. Ideally, there would be a way to store customized labels while keeping a relation to the building data to allow for the later addition of other types of annotations.

One conceivable approach is to store and edit the annotations in a document or database separate from the CityGML file. If we still wanted semantic integration of these annotations with the information from the main file, it would be possible to place (or use existing) identifiers on each XML-entry that is annotated. In many cases, this would already require some modification of the original file, and it would be harder to guarantee the consistency and integrity of both the CityGML and the annotation data set.

A different approach, and one for which the CityGML standard offers a toolset, is adding the annotations directly into the XML-structures of CityGML itself. This is possible by using an Application Domain Extension (ADE). These extensions provide a way to insert domain-specific data into CityGML, without making the city objects themselves non-compliant with the standard. Validation tools can simply ignore entries that were marked as part of an ADE. An example of such an extension is the Noise ADE [6], which is used for storing information pertinent to urban noise propagation like roof and wall materials. This approach would keep all semantic and geometric information for both buildings and annotations in one document and avoid any cross-document references.

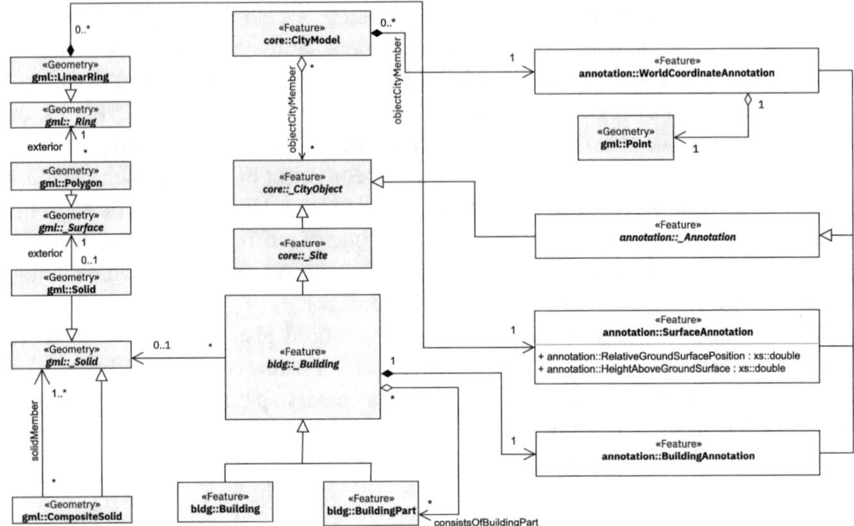

Fig. 1 The ADE allows establishing a topological and semantic relation between CityGML objects and annotations. Uncolored classes belong to the developed ADE, green classes are related to the GML3 standard, blue classes to the CityGML core classes, and yellow classes to the CityGML building module

Figure 1 shows a UML (Unified Modeling Language) class diagram that describes how we integrate our ADE into the building-level of a CityGML file. *BuildingAnnotations* establish a direct semantic reference to whole buildings, *SurfaceAnnotations* a semantic reference to specific building surfaces, while *WorldCoordinateAnnotations* only reference the whole city model. Note that the CityGML standard allows nesting of buildings and parts of buildings into each other. Due to the absence of a clear guideline for subdivision, individual consideration is necessary to either classify ancillary buildings as independent structures or to group them with the superordinate one for *BuildingAnnotations*.

This integrated approach allows us to anchor annotations in the real world by utilizing the topology of the CityGML elements without explicit coordinate specification. The three annotation types each have different anchoring strategies.

BuildingAnnotations are located above buildings. As shown in Fig. 2, we use the centroid of gravity (**C**) of the bounding box (green) as the base of the anchor point (**A**). Depending on the subdivision of complex buildings (parts) into separate nested units in the CityGML data, it may be preferable to start with the centroid of gravity of the combined ground surfaces instead. To derive the height coordinate of **A**, we added an offset to the highest building point (**H**) of the building (part), which increases the visibility of the annotation when viewed from below the roof surface. We have omitted a complex user-location dependent dynamic adjustment of the offset between the highest building point and the anchor point, to keep the calculation simple.

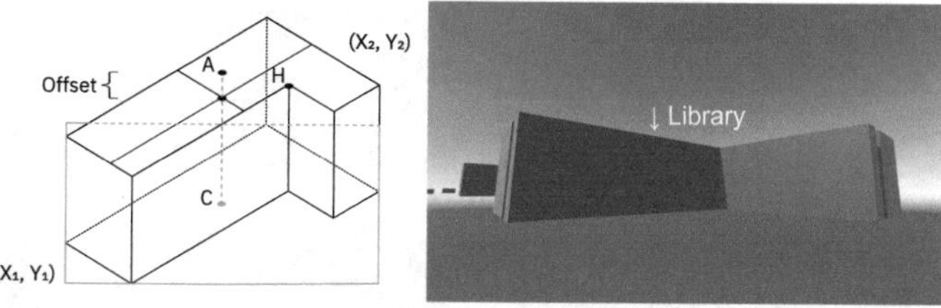

Fig. 2 Up to one BuildingAnnotation can be associated with each building

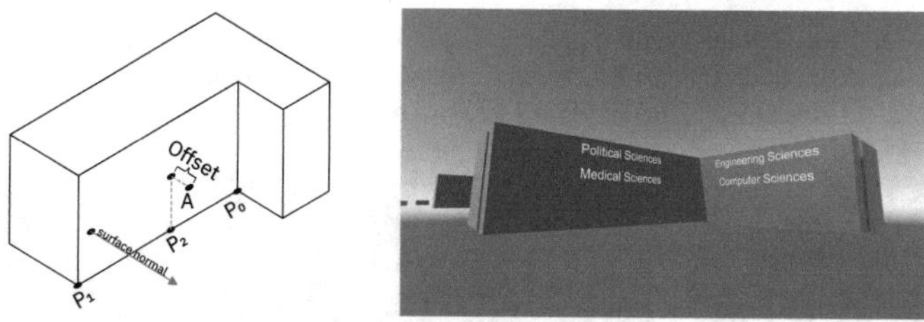

Fig. 3 SurfaceAnnotations are positioned on building surfaces

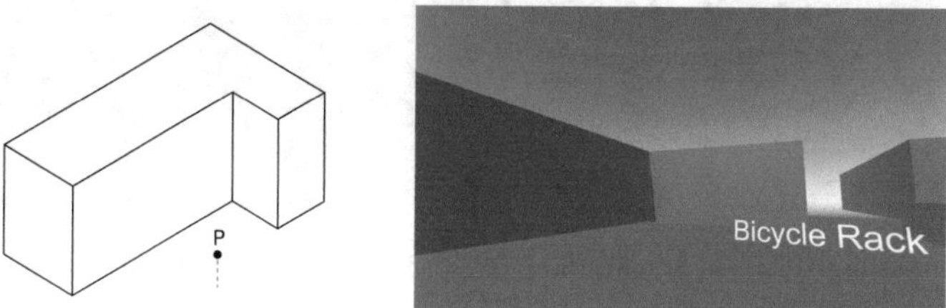

Fig. 4 WorldCoordinateAnnotations are freely placed in 3D space using ETRS89/UTM coordinates

To place *SurfaceAnnotations* (Fig. 3), we use two adjacent polygon points (**P0** and **P1**), since the polygons of building surfaces do not generally need to be rectangular. By specifying the relative position between two points in the form of a value between 0 and 1, we position point **P2** in between the two given polygon points. **P2** is centered if a value of 0.5 is selected. A parameter for height above the baseline, also defined in the ADE, enables positioning of the annotation perpendicular to the surface normal and the vector $\overrightarrow{\mathbf{P1P0}}$. A minor offset parallel to the surface normal ensures visibility by placing the annotation in front of the surface and results in the final anchor point **A** of the surface annotation.

Lastly, to enable annotations that are not building related, ETRS89/UTM coordinates locate *WorldCoordinateAnnotations*, as shown in Fig. 4. While these three annotation anchoring methods are the way we integrate the labels into CityGML, we also need to make them work with Unity. In our prototype, we use a simple 2D text component. However, even the more complex types of annotations like images or videos are usually in 2D, and could thus be handled by Unity's LayoutGroup system.

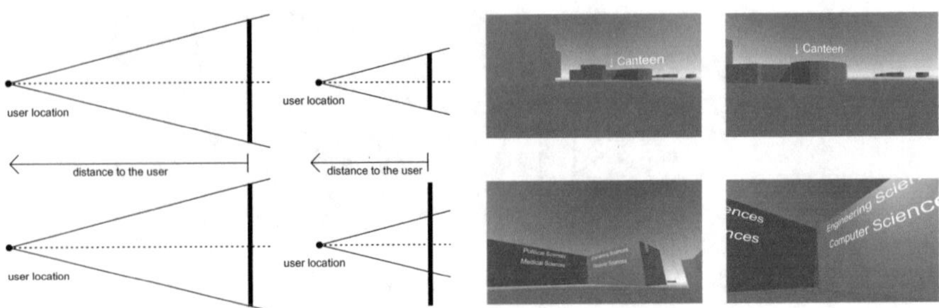

Fig. 5 By location-dependent scaling, the annotation keeps a preset size at the target device (top) while its size changes linearly with the user's distance to the annotation (bottom). In case a fixed size in the real world is chosen, the displayed size changes with viewing distance

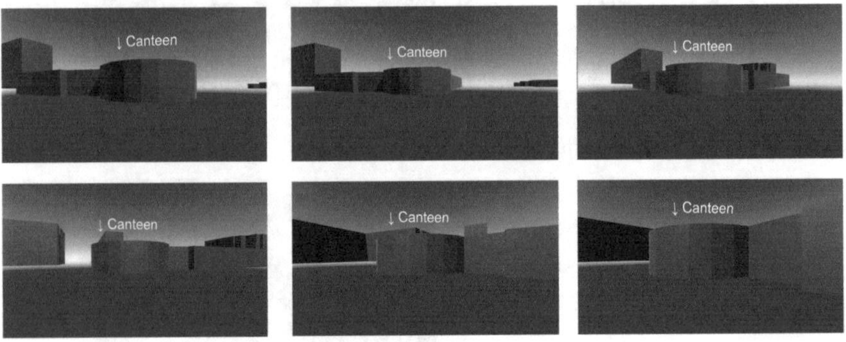

Fig. 6 Besides a fixed alignment, annotations oriented towards the target device are reasonable, as this provides a comprehensive overview of every perspective

Regardless of the annotation's content, proper scaling is essential for an appealing presentation on each target device. Depending on the semantic context, it may be useful to specify the annotation size in three-dimensional space, statically or dynamically adjusted to the distance of the user. We define this behavior through a boolean attribute for each annotation in the ADE.

If an annotation is set to adjust dynamically, it will scale linearly depending on the distance to the viewer and thus change its size in 3D space (Fig. 5). In case the annotation content provides a specific scaling, or a surface annotation should not exceed the size of the associated surface, the annotation has a constant appearance in 3D space, but changes with distance in screen space.

Since the user's perspective on annotations changes during runtime, they orient permanently towards the user, as long as the semantic connection to CityGML elements does not specify an alignment (Fig. 6). The latter is the case only for SurfaceAnnotations, which are oriented orthogonally to the normal of the associated surface.

2.4 Processing the Data

For the use of CityGML data in game engines, it is common to perform a conversion into an FBX model or a COLLADA 3D model before compile-time. This step eliminates the need for a runtime importer that transforms CityGML's boundary representation into a polygon mesh that can be rendered by the engine. However, 3D exchange formats either do not have capabilities for storing the kinds of contextual information and meta-data present in CityGML documents or would require another parsing step to translate them. As we use data sets enriched by the ADE implementation, we need full access to the topological and semantic relations between the CityGML objects at runtime. Although it is not part of this prototype, this would also allow better editing of annotations from within the app. Another advantage of runtime import is that it is easy to load the relevant parts of the scene selectively. Because 3D exchange formats generally do not have geospatial capabilities, a custom loading strategy would have to be implemented (e.g., sorting all models into georeferenced tiles), while directly loading structures from a geospatial format like CityGML can be facilitated through a simple spatial buffer query to a database.

The size of the dataset can vary significantly depending on its composition, level of detail, and spatial coverage, requiring individual assessment for each use case. The CityGML data set used for evaluating our prototype covered a district-sized area of the city and contained a combination of LoD 1 and LoD 2 objects. Its total size was about 175 MB. To have the data locally available on the end device, we initialize an SQLite database with the CityGML data before compiling and integrate the database file into the app.

To import the CityGML data into the database, it is necessary to first deserialize it. Blut et al. [3] evaluate the use of different parsing approaches provided by Java libraries in terms of their memory and time consumption and recommend pull parsers for deserializing CityGML files. Using this approach supplies a high degree of flexibility in controlling sequential reading. We use the .Net Standard 2.0 framework's XMLReader provided by Unity because it is highly performant, and only sequential unidirectional reading is required.

To enable spatial queries based on the user's location with reasonable runtime cost, we implement a custom spatial indexing strategy. For this, we determine two coordinates spanning a bounding box for each building as it is inserted into the database.

AR Foundation's image tracking capabilities allow us to anchor the virtual objects in the physical world at runtime. This makes using image targets effortless—any arbitrary image can be included in the project and be used as a marker image, as long as it has enough visual features to track well. This means that in certain situations, images that already exist in the environment (e.g. posters) could be used for tracking, as long as they can be georeferenced and included in the Unity project ahead of time.

Once one of these image targets is recognized, a spatial query with the target coordinates is made to the database. The query then returns the WorldCoordinateAnnotations, the building geometries, and their associated BuildingAnnotations, as well as the SurfaceAnnotations within a predefined distance along the longitudinal and latitudinal axis. During this process, an on-the-fly conversion of ETRS89/UTM coordinates to the local Unity coordinate system around the detected target is performed, in which only each point's positional difference to the image target is kept. The direct use of a projected coordinate system like ETRS89/UTM is not possible, as Unity's positioning is handled through float vector values with less significant digits than required to represent geospatial coordinates.

While the label data consists only of positioning data and content, the building geometries are explicitly described by vertices and surfaces. Since Unity composes complex surfaces of triangles, a decomposition of each surface of the buildings through triangulation is necessary to describe their geometries unambiguously. There are many available algorithms and libraries for triangulation, however they often assume coordinate values in 2D space. For these cases, it is important to note that CityGML restricts its surfaces to be planar [6], so vertices of a surface are all in one plane in 3D space by definition. We can thus achieve a dimensional reduction by rotating a temporary copy of the surface's polygon points into the horizontal x-y plane. Utilizing Unity's quaternions and the surface's normal vector, the execution is simple and performant. The determined x- and y-coordinates then allow us to perform triangulation in 3D space.

2.5 Occlusion Management

To describe surfaces distinctly, game engines use meshes that combine the initial vertices and the calculated triangles with a material component. Materials consist of information such as textures, shaders, and more to specify how a surface should appear after rendering. However, because the building geometries are used only for occlusion, we just need a shader and no textures. Utilizing Unity's declarative language called ShaderLab, Listing 15.1 shows the shader used for the building mesh's material.

Listing 15.1 The building shader utilized for depth masking, based on [4].

```
1 Shader "DepthMask" {
2     SubShader {
3         Tags {"Queue" = "Geometry"}
4         Pass {
5             ZWrite On
6             ColorMask 0
7         }
8     }
9 }
```

The game engine uses the Z-Buffer method for occlusion calculation, which determines which elements of a 3D scene to incorporate into the color value calculation. Due to the inclusion in the depth buffer (line 5), objects located behind a building are no longer taken into account in the rendering process. To ensure that the occlud-

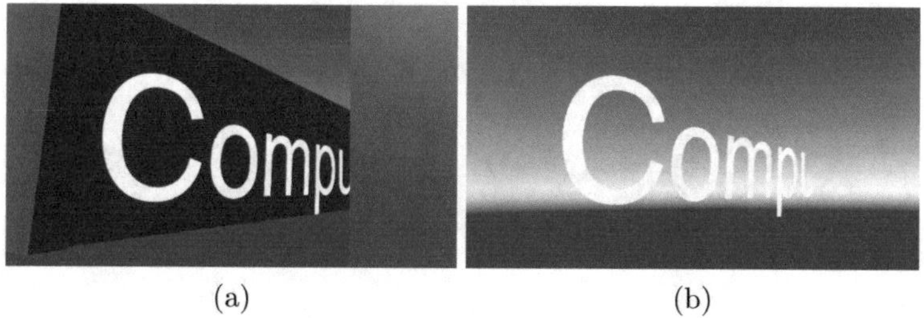

Fig. 7 A SurfaceAnnotation partially occluded by a building using a material with a conventional shader (a) and with the shader predefined in Listing 15.1 for masking purpose with all color channels deactivated (b)

ing building itself is not rendered, all color channels (line 6) are deactivated. This approach allows the implementation of an effective occlusion by buildings without blocking the real-world buildings (Fig. 7).

3 Results and Conclusions

At the end of this process stands a working prototype that shows that an ADE-based annotation approach is feasible and that the same CityGML data used to store and align the annotations can also be utilized to occlude those same annotations. The visual results are shown in four examples in Fig. 8. The described implementation yields a performance that is real-time capable even on mobile devices. For a scene with 62 buildings (2008 vertices) the prototype runs at an average of 59 frames per second (FPS) on an *iPhone 7 Plus*, with infrequent drops to about 30 FPS. On a Samsung *Galaxy S8* the same scene runs at an average of 33 FPS, though it should be noted that this lower performance is already present before the buildings are loaded. On the iPhone the time from successful marker detection to appearance of the labels takes up three seconds, on the Samsung device twelve seconds, likely due to a more efficient SQLite implementation internal to iOS.

Meanwhile, using Unity and *AR Foundation* ensures that the implementation works cross-platform and without any requirement for calibration by the user. If tracking is set up properly, the annotations appear perspectively correct and at the right depth, both from near and far distances. The tracking is stable enough so that there is no apparent jittering even at long distances.

Any standard-compliant CityGML data set of LoD 1 or 2 can be imported into the database and used without recompilation, as long as one has access to the device's file system. Of course, system-related limitations like memory or network availability need to be taken into account when importing new data. However, the image targets still need to be set at compile time. For a production version of the application, it would be possible to reassign their coordinates on the fly, e.g., with a server-based approach.

Fig. 8 **a** A building with a BuildingAnnotation and SurfaceAnnotations that describe its individual floors. **b** A SurfaceAnnotation. **c** A SurfaceAnnotation partially hidden by another building. **d** An imprecise localization by the target approach in combination with drift leads to an improper alignment of virtual objects with the real world

One of the few technical obstacles is the lack of a spatial database extension. Having a spatial DBMS available would likely make the prototype faster and allow for better handling of data at runtime. It would also allow for more heterogeneous data sources, as handling different coordinate reference systems would become easier or could even be automated. However, the more specialized the DBMS plug-in becomes, the less likely it is to work on a wide variety of target devices. One way around these compatibility problems would be to allow for both local and server-side databases to be used. With sufficient network speeds, a server-side database would have few of the restrictions of a local database. Many city models could be stored at once and at many different LoDs, and the whole structure of the database could be geared towards handling CityGML and ADE data, as for example shown by Yao and Kolbe [14].

Meanwhile, another issue lies in the CityGML data itself: the handling of floors, roofs, and building height. Depending on the chosen LoD, building objects contain several pieces of height information: the measured height, as well as the 3D coordinates of the floor and roof vertices. The measured height corresponds to the distance between the lowest intersection point and the highest point of the roof. Therefore, it does not necessarily describe the actual building height, as parts of them may be located below the terrain. However, the floor vertices' characteristics depend heavily on the underlying data source or used surveying method, and could even lie above the terrain, notably if no cellars are registered. Furthermore, CityGML data sets often do not include digital elevation models (DEM) or ones with insufficient accuracy for proper intersecting with building geometries. While for LoD 1 and LoD 2 all surfaces are specified by 3D coordinates, for LoD 0 only the base surface, the roof surface or both are available. If only one of the two surfaces is known, the measured height attribute of the building can be used to derive a prismatic extrusion solid similar to the LoD 1 representation. In case the building height and the measured height do not coincide, either the correct floor or roof placement cannot always be guaranteed.

These issues also call into question the usefulness of the detailed roof shapes, as included in building models from LoD 2 onwards, for our approach. Complex roofs might allow for more accurate positioning of BuildingAnnotations or SurfaceAnnotations specifically related to the roof structure, but the height inaccuracies may very quickly throw off the alignment. This is only compounded by the fact that roof shapes are not always derived from accurate, pre-existing building information models, but can also be estimated from a limited number of roof shape classes or from aerial laser scanning data. It may not be advisable to depend on roof accuracy for the general case, instead annotations should be placed on the exterior walls of buildings where possible. Thus, LoD 1 buildings are sufficient for AR-labeling, except for particular cases like landmark-buildings with complex roof shapes, like churches.

As we follow a purely data-driven approach, there is also currently no radiometric correction or consideration of background color, which means that bright white backgrounds can make the labels hard to read, as demonstrated in Fig. 8a. Additionally, BuildingAnnotation might unintentionally become occluded by their associated building, and SurfaceAnnotations can be hard to read from more oblique angles. Furthermore, the presented occlusion mechanism can't prevent information cluttering if too many annotations are placed in front of or on the same building surface.

If the initial positioning is achieved at sufficient accuracy, both computer-vision-based methods and our method are powerful enough to model basic occlusions with some minor alignment errors around object edges. We expect computer vision methods to perform better in the near field. They can incorporate objects that are not found in the CityGML data, like the trees visible in Fig. 8a, while our approach can more easily handle long-distance occlusions, like on large buildings visible over the rooftops. Since ARCore and ARKit increasingly provide computer-vision-based occlusion functionality, which is also accessible through the AR Foundation framework, we would expect a combination of both mechanisms to yield the best results in the future.

While the target-based localization approach was only implemented to allow for quick and controlled testing, some conclusions can be drawn. Firstly, the non-camera based method of using the GNSS position was found insufficient during a preliminary evaluation, as even in ideal conditions, the alignment would not be precise enough to allow for proper occlusions. Especially building corners are problematic, as shown in Fig. 8d, and require a positioning accuracy of below one meter for correct occlusion. In most environments, the image target approach allows for accurate placement of the buildings and labels, as long as the user does not move too far away from the targets or performs rapid movements. For areas more extensive than a single room, multiple image targets may be necessary to stabilize tracking.

4 Future Research

While Unity enables cross-platform deployment, the prototype is currently only evaluated on smartphone devices. It remains to be investigated whether all the considerations made here hold true for wearable devices like AR HMDs and whether the performance can be sufficient in these cases.

A more fully-featured combination of ADE and the Unity layout-groups would allow for complex annotations, including videos, images, nested layout groups, and extensive interactions beyond location-dependent orientation and scaling. Annotations could unfold once the user points at them or their associated building, or even transform into a standard screen-space 2D layout akin to a website. Advanced, interactive masking and clutter-avoidance strategies could be employed wherever occlusion might not be enough, like in inner-city commercial areas.

Another important note is that, despite its increasing dissemination, CityGML data is not freely available for many regions. For these cases our considerations show that even LoD 0 data can be sufficient for proper annotation and masking. This is equivalent to the level of building geometries usually provided in OSM. However, the accuracy of the occlusion may be reduced, and storing and managing annotations would likely require a more complex and less topologically consistent approach than the ADE solution presented here.

In this paper we also focused mostly on the technical aspects of annotations in AR. Once an application based on this approach is implemented for a specific purpose, the appearance and placement of annotations needs to be considered in terms of usability and user satisfaction. On the software side this would for example include some form of dynamic text coloring for proper legibility and on the design side a targeted user evaluation - an AR system for expert users could look very different than a system for the general public.

In terms of use cases, we only considered the case of pre-authored content that is then delivered to users through the app. However, our semantics and topology-preserving method would also allow for users to create annotations for buildings with their target device and subsequently transmit them back to planners. Thus, the use of an ADE not only facilitates data exchange but can also contribute to the serviceability of the data[1] set.

References

1. Azuma, R., Furmanski, C.: Evaluating label placement for augmented reality view management. In: The Second IEEE and ACM International Symposium on Mixed and Augmented Reality (ISMAR), 2003. Proceedings, pp. 66–75. IEEE, Tokyo, Japan (2003). https://doi.org/10.1109/ISMAR.2003.1240689

[1]The code for this paper is published at: http://github.com/Postert/CityGML-AR.

2. Blut, C., Blankenbach, J.: Three-dimensional CityGML building models in mobile augmented reality: a smartphone-based pose tracking system. Int. J. Digital Earth, pp. 1–20 (2020). https://doi.org/10.1080/17538947.2020.1733680
3. Blut, C., Blut, T., Blankenbach, J.: CityGML goes mobile: application of large 3d CityGML models on smartphones. Int. J. Digital Earth **12**(1), 25–42 (2019). https://doi.org/10.1080/17538947.2017.1404150
4. Carter, N., Brauer, D.: Depthmask. Unify Community Wiki (2018). http://wiki.unity3d.com/index.php?title=DepthMask. Accessed 30 Apr 2020
5. Christensen, J., Marks, J., Shieber, S.: An empirical study of algorithms for point-feature label placement. ACM Trans. Graphics (TOG) **14**(3), 203–232 (1995). https://doi.org/10.1145/212332.212334
6. Gröger, G., Kolbe, T.H., Nagel, C., Häfele, K.H.: OGC City Geography Markup Language (CityGML) Encoding Standard, 2.0.0 edn. Open Geospatial Consortium (2012)
7. Hirzer, M., Arth, C., Roth, P., Lepetit, V.: Efficient 3d tracking in urban environments with semantic segmentation. In: Proceedings of the British Machine Vision Conference (BMVC), London, UK (2017)
8. Langlotz, T., Nguyen, T., Schmalstieg, D., Grasset, R.: Next-generation augmented reality browsers: rich, seamless, and adaptive. Proc. IEEE **102**(2), 155–169 (2014). https://doi.org/10.1109/JPROC.2013.2294255
9. Löwner, M.O., Gröger, G., Benner, J., Biljecki, F., Nagel, C.: Proposal for a new lod and multi-representation concept for CityGML. In: ISPRS Annals of Photogrammetry, Remote Sensing & Spatial Information Sciences, vol. IV-2/W1, pp. 3–12 (2016). https://doi.org/10.5194/isprs-annals-IV-2-W1-3-2016
10. Löwner, M.O., Casper, E., Becker, T., Benner, J., Gröger, G., Gruber, U., Häfele, K.H., Kaden, R., Schlüter, S.: Citygml 2.0 – ein internationaler standard für 3d-stadtmodelle, teil 2: Citygml in der praxis. Zeitschrift für Geodäsie, Geoinformation und Landmanagement **2**(2013), 131–143 (2013)
11. McNamara, A., Boyd, K., George, J., Jones, W., Oh, S., Suther, A.: Information placement in virtual reality. In: Teather, R., Itoh, Y., Gabbard, J. (eds.) Proceedings, 26th IEEE Conference on Virtual Reality and 3D User Interfaces, pp. 1765–1769. IEEE, Osaka, Japan (2019). https://doi.org/10.1109/VR.2019.8797891
12. Reitmayr, G., Drummond, T.W.: Going out: robust model-based tracking for outdoor augmented reality. In: 2006 IEEE/ACM International Symposium on Mixed and Augmented Reality, pp. 109–118. IEEE, Santa Barbard, CA, USA (2006). https://doi.org/10.1109/ISMAR.2006.297801
13. Schöps, T., Sattler, T., Häne, C., Pollefeys, M.: 3d modeling on the go: Interactive 3d reconstruction of large-scale scenes on mobile devices. In: 2015 International Conference on 3D Vision, pp. 291–299. IEEE, Lyon, France (2015). https://doi.org/10.1109/3DV.2015.40
14. Yao, Z., Kolbe, T.H.: Dynamically extending spatial databases to support citygml application domain extensions using graph transformations. In: T.P. Kersten (ed.) Kulturelles Erbe erfassen und bewahren - Von der Dokumentation zum virtuellen Rundgang, 37. Wissenschaftlich-Technische Jahrestagung der DGPF, *Publikationen der Deutschen Gesellschaft für Photogrammetrie, Fernerkundung und Geoinformation (DGPF) e.V*, vol. 26, pp. 316–331. Deutsche Gesellschaft für Photogrammetrie, Fernerkundung und Geoinformation e.V, Würzburg (2017)
15. Zama Ramirez, P., Poggi, M., Tosi, F., Mattoccia, S., Di Stefano, L.: Geometry meets semantics for semi-supervised monocular depth estimation. In: Jawahar, C.V., Li, H., Mori, G., Schindler, K. (eds.) Computer Vision – ACCV 2018, pp. 298–313. Springer International Publishing, Cham (2019). https://doi.org/10.1007/978-3-030-20893-6_19

Physical Environments

Can Animal Manure Be Used to Increase Soil Organic Carbon Stocks in the Mediterranean as a Mitigation Climate Change Strategy?

Andreas Kamilaris, Immaculada Funes Mesa, Robert Savé, Felicidad De Herralde, and Francesc X. Prenafeta-Boldú

Abstract Soil organic carbon (SOC) plays an important role on improving soil conditions and soil functions. Increasing land use changes have induced an important decline of SOC content at global scale. Increasing SOC in agricultural soils has been proposed as a strategy to mitigate climate change. Animal manure has the characteristic of enriching SOC, when applied to crop fields, while, in parallel, it could constitute a natural fertilizer for the crops. In this paper, a simulation is performed using the area of Catalonia, Spain as a case study for the characteristic low SOC in the Mediterranean, to examine whether animal manure can improve substantially the SOC of agricultural fields, when applied as organic fertilizers. Our results show that the policy goals of the 4 × 1000 strategy can be achieved only partially by using manure transported to the fields. This implies that the proposed approach needs to be combined with other strategies.

Keywords Soil organic carbon stock · Animal manure · Climate change

1 Introduction

Soil organic carbon (SOC) plays a crucial role on physical, chemical and biological soil conditions and consequently on soil functions such as sustaining and enhancing crop yields due to the increase in water and nutrients storage and availability for plants [1]. However, an important decline of SOC content has been observed at the global scale because of land use changes, like deforestation in favour of cultivation and intensive agricultural practices [2]. SOC depletion is basically promoted by changes in soil temperature and texture, moisture regimes, soil disturbance and erosion [1].

A. Kamilaris (✉)
Research Centre On Interactive Media, Smart Systems and Emerging Technologies (RISE), Nicosia, Cyprus
e-mail: a.kamilaris@rise.org.cy

Department of Computer Science, University of Twente, Enschede, The Netherlands

I. F. Mesa · R. Savé · F. De Herralde · F. X. Prenafeta-Boldú
Institute of Agrifood Research and Technology, Barcelona, Spain

A. Kamilaris et al. (eds.), *Advances and New Trends in Environmental Informatics*, Progress in IS, https://doi.org/10.1007/978-3-030-61969-5_16

Estimates indicate that SOC is being lost globally at an annual rate equivalent to 10–20% of the total global carbon dioxide emissions. Most soil landscapes in the southern and eastern parts of the Mediterranean basin, independently from the cropping systems, are subject to SOC losses which are mainly attributable to relatively high temperatures and soil erosion. In the northern part of the basin, the issue of low SOC stocks is of particular concern in perennial systems such as orchards and vineyards [3], which play an important socioeconomic role in southern Europe.

Recent data showed that bare soils, vineyards and orchards in Europe are prone to erosion (10–20 tonnes ha^{-1} yr^{-1}), while cropland and fallow show smaller soil losses (6.5 and 5.8 tonnes ha^{-1} yr^{-1}) largely because the latter occupy land with little or no slope. Grasslands and the associated livestock rearing are of limited extent in Mediterranean regions, so the accumulation of SOC associated with such land uses is severely restricted. Overgrazing is a potential threat though. In addition, wildfires, which are rather common in the Mediterranean, can also have a negative impact on SOC, but they normally affect forests and rangelands and are thus of limited concern in cultivated agro-ecosystems. In the particular case of Spain, there was a continued decline in SOC during the twentieth century (cropland SOC levels in 2008 were 17% below their 1933 peak) and these SOC trends were driven by historical changes in land uses, management practices and climate [4]. In this sense, the FOOD chapter of the first MedECC report focused on the loss of soil organic carbon (SOC) [5]. Fortunately, certain soil management practices may significantly influence the capacity of the soil to sequester SOC in agricultural soils [1]. Several broad strategies have been suggested to increase or maintain soil fertility (water storage, nutrients source, biodiversity preservation, etc.) since the very start of agriculture, although the recent intensification has been distorting soil fertility. The main strategies to increase SOC contents and fertility in Mediterranean soils include reduced soil tillage, crop rotations, cover crops and the introduction of new crop varieties or cultures, among others. In addition, synergies and trade-offs between adaptation and mitigation strategies must be considered for an enhanced resilience of agro-systems, particularly in regions highly impacted by climate change such as the Mediterranean [6, 7].

In this context, the international initiative "*4 per 1000 Soils for Food Security and Climate*" was launched by France on 1st December 2015 at the 21st Conference of the Parties to the United Nations Framework Convention on Climate Change (COP21) and validated on November 16 in Marrakech (COP22). The "4 per 1000" (4 × 1000) initiative consists of federating all voluntary stakeholders of the public and private sectors (national governments, local and regional governments, companies, trade organisations, NGOs, research facilities, etc.) under the framework of the Lima-Paris Action Plan (LPAP). The aim of the initiative is to demonstrate that agriculture, and in particular agricultural soils, can play a crucial role where food security and climate change are concerned by increasing soil organic matter stocks by 4 per 1000 (or 0.4%) per year as a compensation for the global emissions of greenhouse gases by anthropogenic sources [8]. This strategy has been included within the agronomic practices focused on the mitigation and adaptation of agriculture to climate change in Catalonia (Northeast of Spain). Supported by solid scientific documentation, this initiative invites all participants and stakeholders to state and implement specific practices aimed at enhancing soil carbon storage and the type of practices to achieve this (e.g. agroecology, agroforestry, conservation agriculture, landscape management, etc.). The ambition of the initiative is to encourage stakeholders to transition towards a productive, highly resilient agriculture, based on the appropriate management of lands and soils, creating jobs and incomes hence ensuring sustainable development.

The present study assesses the application of organic matter in agricultural soils as an option of carbon sequestration strategy at the regional scale, considering as well the problems associated to organic matter characteristics, transportation and application in a global carbon and economic footprint have also been considered in this assessment. This will allow to assess the feasibility of reaching the 4×1000 target by means of animal manure used as agricultural fertilizer.

2 Related Work

Surplus manure from livestock has been used in the past as fertilizer in crop fields. An approach for the transportation of manure beyond individual farms for nutrient utilization was proposed in [9], focusing on animal manure distribution in Michigan.

Teira-Esmatges and Flotats [10] proposed a methodology to distribute manure at a regional and municipal scale in an agronomically correct way, i.e. by balancing manure application based on territorial crop needs, as well as on predictions of future needs and availability considering changes in land use. ValorE [11] is a GIS-based decision support system for livestock manure management, with a small case study performed at a municipality level in the Lombardy region, northern Italy, indicating the feasibility of manure transfer. Other researchers proposed approaches to select sites for safe application of animal manure as fertilizer to agricultural land [12, 13]. Site suitability maps have been created using a GIS-based model in the Netherlands and in Queensland, Australia respectively.

Several studies reported agricultural management practices that are found to be beneficial to sequester carbon from various regions in the world (see the literature review and listed sequestration rates in [8]. Addition of organic amendments to the soil is one of the more reported management practices to increase SOC, along with other agricultural practices [14–18] such us reduced tillage [19, 20], crop residue incorporation [20–22], cover crops [23, 24], crop rotation [20, 25] or land use changes [26–28].

Moreover, organic farming [4, 29–31] and conservation agriculture [32] are reported to effectively promote carbon sequestration when compared to conventional agriculture. Most of these recommended practices are supposed to increase SOC by themselves but some studies have highlighted the potential of combining practices to improve sequestration [16–18]. Moreover, there are studies on upscaling the effects of SOC on climate change, land use changes or the application of certain agricultural practices [15, 26, 33]. However, compost or organic amendments large scale estimates for the carbon sequestration potential of compost and organic amendments in agricultural soils are still scarce [18].

3 Methodology

The purpose of this section is to describe how the problem was modelled using the area of Catalonia as a case study.

3.1 Problem Description

Catalonia is one of the European regions with the highest livestock density, according to the agricultural statistics for 2016, provided by the Ministry of Agriculture, Government of Catalonia. The animal census reported numbers of around 9 M pigs, 0.7 M cattle and 75 M poultry in a geographical area of 32,108 square kilometers, hosted in 25 K livestock farms. The overall goal is to solve the problem of how to find the optimal and economically viable way to distribute animal manure in order to fulfil agricultural fertilization needs in one hand and to increase carbon stock of these fields on the other hand. To simplify the problem, the geographical area of Catalonia has been divided into a two-dimensional grid, as shown in Fig. 1 (left). In this way, the distances between livestock farms (i.e. original grid cell) and crop fields (e.g. destination grid cell) are easier to compute, considering straight-line grid cell Manhattan distance as the metric to use and not actual real distance through the

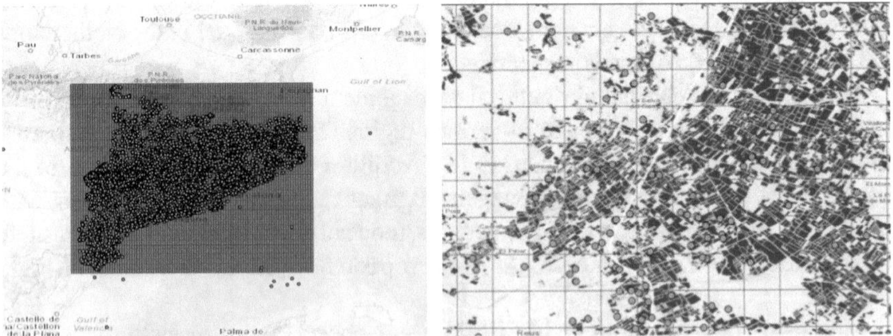

Fig. 1 Division of the territory of Catalonia in cells of 1 km^2. each (left). Demonstration of livestock farms and crop fields at grid cells in a dense agricultural area of the region (right). This is a zoom of the map shown on the left. Livestock farms are shown as brown circles, and crop fields as blue polygons

existing transportation network. Each crop field and livestock farm has been assigned to the grid cell where the farm is physically located.

3.2 *Data Collection and Pre-processing*

Details about livestock farms (i.e. animal types and census, location etc.) have been provided by the Department of Agriculture of the Government of Catalonia for the year 2016, after signing a confidentiality agreement. Details about crop fields (i.e. crop type, hectares, irrigation method, location etc.) have been downloaded from the website of the Department [34], for the year 2015. More than 20 K crop fields have been recorded.

For every livestock farm, the yearly amount of manure produced and its equivalent in nitrogen as fertilizer have been calculated, depending on the type and number of animals on the farm, based on the IPCC guidelines (TIER1) [35]. Similarly, for every crop field, the yearly needs in nitrogen have been computed, depending on the crop type and total hectares of land, according to the Nitrate Directive [36]. The estimated total fertilizer needs of crop fields (i.e. 88 K tons of nitrogen) were lower than the availability of nitrogen from animal manure (i.e. 116 K tons of nitrogen) [37, 38]. This means that the produced amount of manure/nitrogen from livestock agriculture has the potential to completely satisfy the total needs of crop farms. Further, we needed to know the existing soil organic content (SOC) in soils around Catalonia and to associate this with the existing crop fields. This information was retrieved from the SOC stock baseline map for Catalonia, published in [39] and downloaded from the Institut Cartogràfic i Geològic de Catalunya (ICGC) [40]. The correlation with crop fields was performed in ESRI ArcGIS. Finally, it was necessary to estimate the carbon content by applying animal manure to the Catalonian soils. Table 1 shows the relevant parameters used to estimate the kilograms of carbon contained in animal manure in terms of volume (m^3) or weight (kg). The values are taken from [41–43].

3.3 *Modelling*

The total area of Catalonia has been divided into 74,970 grid cells, each representing a 1×1 square kilometer of physical land. Every cell has a unique ID and (x,y) coordinates, ranging between [1315] for the x coordinate and [1238] for the y coordinate.

For each grid cell, we are aware of the crop and livestock farms located inside that cell, the manure/nitrogen production (i.e. from the livestock farms) and the needs in nitrogen (i.e. of the crop fields). Moreover, some crop farms in Catalonia fall within *nitrate vulnerable zones*. Inside these zones, only a maximum of *170 kg/ha* (i.e. kilos/hectare) of nitrogen from manure can be applied. We assumed that each transport vehicle (i.e. truck) used for the transfer of manure has a limited capacity

Table 1 Equivalence table of relevant parameters for each animal fertilizer type to estimate carbon contained in animal manure

	Density[1] (ton/m3)	Relative unit	Dry matter[2] (kg)	Organic matter[2] (kg)	Total organic carbon (kg C/m^3)		kg C/kg fertilizer
Fertilizer type					Range	Average value	Average value
Pig slurry (fattening)	1.050	Per 1 m^3	70	40–50	20.86–26.08	23.47	0.0224
Pig slurry (maternity)	1.024	Per 1 m^3	50	20–30	10.43–15.65	13.04	0.0127
Manure (bovine)	0.750	Per 1 ton	210	120–150	62.60–7825	70.42	0.07042
Manure (poultry)	0.850	Per 1 ton	600	370–420	193.01–219.09	206.05	0.20605
Compost (bovine manure)[3]	0.720	Per 1 ton	508	274.8	–	143.35	0.14335
Compost (poultry manure)[3]	0.720	Per 1 ton	754	418.4	–	218.26	0.21826

of 20 cubic meters to transfer manure/nitrogen. The allowed periods of the year when fertilizer can be applied on the land (depending whether the crop falls within a vulnerable area or not) based on the Directive 153/2019 of the Government of Catalonia [44], were also considered. A uniform monthly production of manure was assumed, not affected by the season.

3.4 Simulation

To solve the manure transfer problem from livestock farms to crop fields in an optimal way, a centralized optimized approach (COA) has been developed [37], based on an algorithm that generalizes and adapts the well-known Dijkstra's algorithm for finding shortest paths [45], together with the use of origin–destination cost matrices as applied in the travelling salesman problem for choosing best routes [46]. COA solves the problem by considering a shortest-path problem on an undirected, non-negative, weighted graph. COA is described in [37]. To use the algorithm within the context of the problem under study, the algorithm has been modified to respect the necessary configurations and constraints, i.e. by modelling the weights of the graph to represent both transport distances and crop farms' nitrogen needs. All combinations of visits to nearby farms (within a wide radius of 100 km) are added to an origin–destination cost matrix, where the most profitable route is selected. In contrary to the typical travelling salesman problem, here the possible stop locations vary depending on which combinations of candidate crop farms maximize the following global objective (GO):

$$GO = \left(NT^{*}0.225^{*}l\right) - \left(TD^{*}0.1827\right) \tag{1}$$

where *NT* is the total nitrogen transferred in kilograms at every *transaction*, and *TD* is the total distance in kilometers covered to transport manure from the livestock farm to the crop field. The parameter l captures the nutrient losses of manure during its storage time, i.e. the time when the manure is stored at the livestock farm until it is transferred to the crop field. A loss of 5% has been considered based on [47].

A decentralized, nature-inspired approach based on an ant colony optimization algorithm has been published by the authors as well [38], but with 8% less performance (in terms of the global objective GO) in comparison to COA.

4 Results

This section presents the findings obtained by solving the problem of manure transport optimization, based on the modelling and the simulations performed in Sect. 3.

Table 2 summarizes the total carbon stored at each different crop type, based on the transfers that took place after running the simulator for a complete one-year duration. A total of 106 thousand tons of carbon is indirectly stored into 10,982 fields of 19 different crop types, spread at a total area of 390 thousand hectares. The mean value of carbon added in each crop field where animal manure was applied was 0.29 kg C/m2 as a consequence of applying mean doses of animal fertilizer around 1.34 kg/m2.

Figure 2 visualizes the results of Table 1 in a heatmap, to highlight the differences in SOC accumulated in different crop types. As Fig. 2 shows, cereals have the most accumulation during the months of August-November. Some worth-mentioning SOC storage occurs also between January–March for forage, July for rapeseed, April for alfalfa and December for sweet fruits and olive trees.

The map in Fig. 3 visualizes the crop fields that satisfy the 4 × 1000 strategy after the first year of the approach under study is applied. These crop fields are depicted

Table 2 Kilograms of carbon (in thousands) stored in different types of crop fields in the hypothetical scenario of transferring animal manure from livestock farms to crop fields, after a year

Crop	Month											
	1	2	3	4	5	6	7	8	9	10	11	12
Alfalfa	0	0	783	3184	1541	102	239	164	0	0	0	0
Corn	1677	1449	1433	209	55	60	39	2	0	0	0	397
Cereals	441	318	235	31	0	0	0	5545	20,443	4386	4035	593
Vegetables	407	161	345	201	107	113	62	57	40	33	37	35
Olive trees	1606	1397	49	21	22	12	5	0	0	0	1	3
Sweet fruits	1271	965	25	14	4	9	0	0	0	0	230	2971
Wood-based crops	790	459	516	310	270	301	157	110	76	55	90	108
Vineyards	535	465	4	3	2	3	0	0	0	0	667	712
Nut trees	971	713	7	3	1	0	0	0	0	0	585	1543
Fallow	0	0	0	0	0	0	0	0	0	0	0	0
Forage	3802	2918	2752	2233	1831	1996	1470	1250	773	725	684	1024
Sunflower	383	270	212	17	10	7	4	0	0	0	0	59
Rapeseed	234	169	27	0	0	0	3060	566	182	181	381	101
Legumes	456	168	186	134	65	94	62	28	12	6	6	16
Other crops	681	439	513	403	330	288	204	173	69	76	83	117
Citruses	162	123	0	0	0	0	0	0	0	0	16	731
Hazelnut	144	141	7	1	1	1	0	0	0	0	168	191
Hemp	30	15	12	4	6	4	4	4	4	4	4	6
Rice	48	47	52	50	53	39	35	32	34	33	29	43
Soya	34	20	19	17	9	5	8	4	6	5	4	2
Camelina	0	0	0	0	0	0	0	0	0	0	0	0

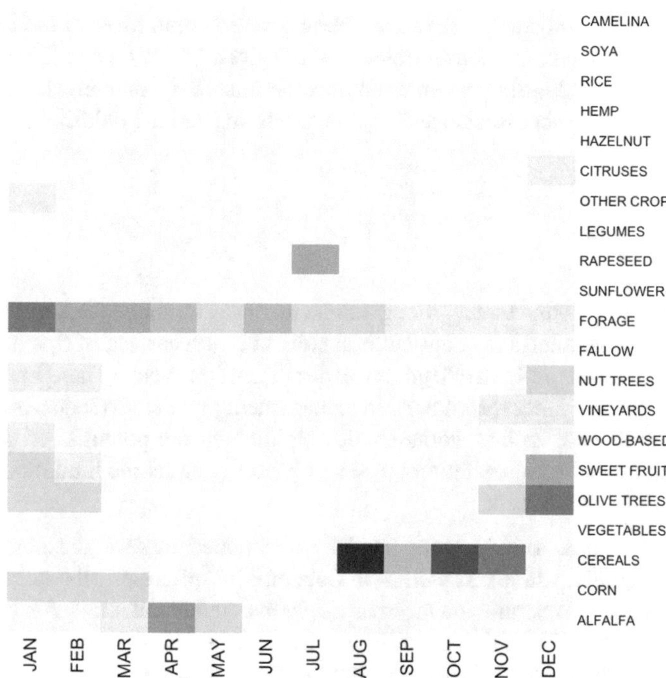

Fig. 2 Heatmap visualizing SOC accumulated in different crop types, in different year months

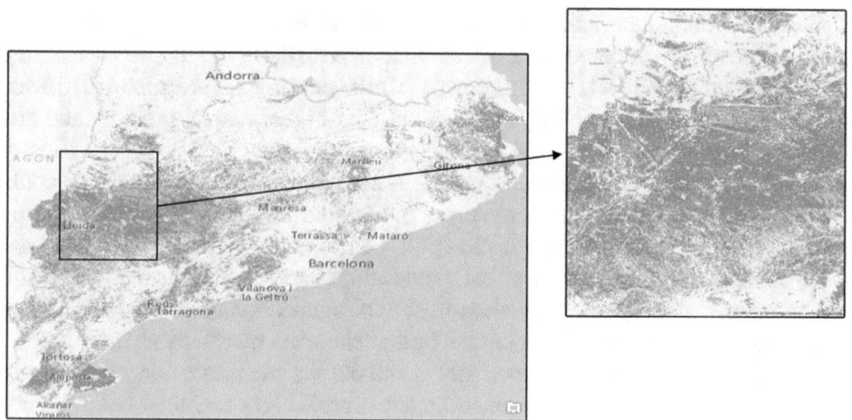

Fig. 3 Map of Catalonia showing the crop fields of the region. The fields satisfying the 4 × 1000 strategy at the first year of the manure transfer program are shown in green color, the rest fields in red color. The region near the city of Lleida is magnified at the right side

in green color, in comparison to the rest shown in red color. Most of the crop fields satisfying the strategy are concentrated in the regions of Lleida and Girona, where livestock farms are highly concentrated thus the transfer of manure is possible and efficient, due to the short distances between the farms and the fields.

5 Discussion

The present study found that applying animal manure as a single strategy to increase SOC stocks in Mediterranean agricultural soils in North eastern of Spain would be not enough to reach the 4 × 1000 goals in a duration of one year.[1] Therefore, it will be necessary to assess other recommended management practices to sequestrate carbon in agricultural soils. In fact, some studies highlighted the potential of combining practices to improve sequestration [16–18]. Our study has some limitations that are important to be mentioned:

(a) SOM for some animal types has been calculated in several studies outside Europe (India, China), as well as in Catalonia, at different soil depths.
(b) The effect of manure management systems (treatment units for pig slurry, compost units for bovine manure) was not considered.
(c) Only the first year of manure application is actually assessed. Long time periods have been considered only naively, without considering weather/climate modelling/forecasting.

Climate variables and soil properties are important drivers of SOC dynamics [39], but have not been considered in the calculations, even in the first-year case of applying the proposed approach. It is widely known that climate variables are important drivers of SOC stock: increasing SOC is associated with higher annual precipitation and lower temperature [48, 49]. Higher temperatures are associated to higher rates of mineralization due to the increase of microbial activity. However, soil moisture could act as the main driver of soil microbiomes in Mediterranean environments, limiting SOC losses by microbial mineralization [50]. Otherwise, a decrease in available soil water content would negatively affect yields and, consequently, the associated soil carbon input. Soil properties can also affect SOC stocks as much as organic carbon is stabilized by means of physical protection or chemical mechanisms [51].

The effects of climate change on global soil carbon stocks are controversial [8]. However, it seems clear from the environmental projections that the Mediterranean is warming at a pace that is 20% faster than the global average, leading to a possible regional increase of 2.2 °C of temperature and a 15% reduction in rainfall by 2040, considering that current policies would still be in place by then. The combination of these weather events will promote a high evapotranspiration and, consequently, an increase in the duration and frequency of droughts [5].

---

[1]Further analysis not shown in this paper shows that the 4 × 1000 goal cannot be reached by more than 60% of the farms even after 10 years.

The obtained results in this study are important for the 4 × 1000 strategy under the prism of our actual agricultural system, because it indicates that the soil can improve the carbon balance in specific farming regions and scenarios (see Fig. 3), but it can only contribute partly. The fact of improving only certain regions could be associated to the concentration of livestock farms in specific areas and to the fact that pig slurry, containing low carbon content (Table 1), is the major animal fertilizer produced in Catalonia. In order to increase the possibilities to achieve the 4 × 1000 strategy and prevent ground water pollution by nitrogen, it would be necessary to apply composted manure that contains higher content of carbon. It is true that soil is a very important carbon sink, but the application of 4 × 1000 must be accomplished according to new integrated models of agriculture, in which sources and sinks of carbon will be closed. The method of transferring manure needs to be combined with other strategies, such as those cited in Sect. 2 (Related work) [14–32]: cover crops, crop rotation, reduced tillage, crop residue incorporation to the soil, among others.

5.1 Future Work

The results of the present study indicate that merely with the application of manure in agricultural soils, the 4 × 1000 strategy could not be reached in a reasonable number of years for all the existing crop fields. Hence, future efforts should focus on assessing the upscaling of different management practices that are proved to sequestered carbon in the soil by themselves and the combination of different practices. Moreover, it is important for future work to consider the effects of:

i. Precipitation/temperature changes using different climate change scenarios;
ii. Land uses changes;
iii. More realistic models to calculate the SOM based on animal manure;
iv. Manure management systems and multiple years; and
v. CO_2 emissions of manure transportation from livestock farm to crop field.

6 Conclusion

Soil organic carbon (SOC) plays an important role in improving soil conditions and soil functions. Increasing land use changes have induced an important decline of SOC content at global scale. Animal manure has the characteristic of increasing SOC, when applied to crop fields, while, in parallel, it constitutes a natural fertilizer for the crops. In this paper, we have developed a large-scale simulation, using the whole area of the region of Catalonia (Spain) as a case study. The goal was to examine whether animal manure can improve substantially the SOC of the Catalan crop fields. Our results showed that the policy goals of Spain can only be achieved partly by using merely manure transported to the fields, thus this strategy needs to be combined with other agricultural practices. We have discussed implications of the findings, together with additional policies, actions and methods required in order to reach the policy goals of Spain in terms of enriching the SOC in Spanish soils, mitigating the climate change effects.

Acknowledgements This research was supported by the CERCA Programme/Generalitat de Catalunya. The assistance of the Catalan Ministry of Agriculture, Livestock, Fisheries and Food is also acknowledged. Francesc Prenafeta-Boldú belongs to the Consolidated Research Group TERRA (ref. 2017 SGR 1290).

Andreas Kamilaris has received funding from the European Union's Horizon 2020 research and innovation programme under grant agreement No 739578 complemented by the Government of the Republic of Cyprus through the Directorate General for European Programmes, Coordination and Development.

References

1. Lal, R., Delgado, J.A., Groffman, P.M., Millar, N., Dell, C., Rotz, A.: Management to mitigate and adapt to climate change. J. Soil Water Conserv. **66**, 276–285 (2011)
2. Smith, P.: Agricultural greenhouse gas mitigation potential globally in Europe and in the UK: what have we learnt in the last 20 years? Glob. Change Biol. **18**, 35–43 (2012)
3. Montanaro, G., Xiloyannis, C., Nuzzo, V., Dichio, B.: Orchard management, soil organic carbon and ecosystem services in Mediterranean fruit tree crops. Sci. Hortic. **217**, 92–101 (2017)
4. Aguilera, E., Guzmán, G.I., Álvaro-Fuentes, J., Infante-Amate, J., García-Ruiz, R., Carranza-Gallego, G., Soto, D., González de Molina, M.: A historical perspective on soil organic carbon in Mediterranean cropland (Spain, 1900–2008). Sci. Total Environ. **621**, 634–648 (2018)
5. Network of Mediterranean Experts on Climate and Environmental Change (MedECC): Risks associated to climate and environmental changes in the Mediterranean region. a preliminary assessment by the MedECC network. Science-policy interface (2019)
6. Sebastia, M.T., Plaixats, J., Lloveras, J., Girona, J., Caiola, N., Savé. R.: Chapter 13, Sistemes agroalimentaris. In: Martin Vide, J. (ed.) Tercer Informe sobre el Canvi Climàtic a Catalunya. Institut d'estudis Catalans i Generalitat de Catalunya, Barcelona, pp. 315–336
7. Vayreda, J., Retana, J., Savé, R., Funes, I., Sebastià MT., Calvo, E., Catalan. J. Batalla, M.: Chapter 3, Balanç de carboni: els embornals a Catalunya. In: Martin Vide, J. (ed.) Tercer Informe sobre el Canvi Climàtic a Catalunya. Institut d'estudis Catalans i Generalitat de Catalunya, Barcelona, pp. 65–92 (2016)
8. Minasny, B., Malone, B.P., McBratney, A.B., Angers, D.A., Arrouays, D., Chambers, A., Chaplot, V., Chen, Z.S., Cheng, K., Das, B.S., Field, D.J., Gimona, A., Hedley, C.B., Hong, S.Y., Mandal, B., Marchant, B.P., Martin, M., McConkey, B.G., Mulder, V.L., O'Rourke, S., Richer-de-Forges, A.C., Odeh, I., Padarian, J., Paustian, K., Pan, G.X., Poggio, L., Savin, I., Stolbovoy, V., Stockmann, U., Sulaeman, Y., Tsui, C.C., Vagen, T.G., van Wesemael, B., Winowiecki, L.: Soil carbon 4 per mille. Geoderma **292**, 59–86 (2017)
9. He, C., Shi, C.: A preliminary analysis of animal manure distribution in Michigan for nutrient utilization. JAWRA J. Am. Water Resour Assoc **34**(6), 1341–1354 (1998)
10. Teira-Esmatges, M.R., Flotats, X.: A method for livestock waste management planning in NE Spain. Waste Manage. **23**(10), 917–932 (2003)
11. Acutis, M., Alfieri, L., Giussani, A., Provolo, G., Di Guardo, A., Colombini, S., Bertoncini, G., Castelnuovo, M., Sali, G., Moschini, M., Sanna, M.: ValorE: an integrated and GIS-based decision support system for livestock manure management in the Lombardy region (northern Italy). Land Use Policy **41**, 149–162 (2014)

12. Van Lanen, H.A.J., Wopereis, F.A.: Computer-captured expert knowledge to evaluate possibilities for injection of slurry from animal manure in the Netherlands. Geoderma **54**(1–4), 107–124 (1992)
13. Basnet, B.B., Apan, A.A., Raine, S.R.: Selecting suitable sites for animal waste application using a raster GIS. Environ. Manage. **28**(4), 519–531 (2001)
14. Chen, Z., Wang, J., Deng, N., Lv, C., Wang, Q., Yu, H., Li, W.: Modeling the effects of farming management practices on soil organic carbon stock at a county-regional scale. CATENA **160**, 76–89 (2018)
15. Bleuler, M., Farina, R., Francaviglia, R., di Bene, C., Napoli, R., Marchetti, A.: Modelling the impacts of different carbon sources on the soil organic carbon stock and CO2 emissions in the Foggia province (Southern Italy). Agric. Syst. **157**, 258–268 (2017)
16. Pardo, G., del Prado, A., Martínez-Mena, M., Bustamante, M.A., Martín, J.A.R., Álvaro-Fuentes, J., Moral, R.: Orchard and horticulture systems in Spanish Mediterranean coastal areas: is there a real possibility to contribute to C sequestration? Agr. Ecosyst. Environ. **238**, 153–167 (2017)
17. Vicente-Vicente, J.L., Garcia-Ruiz, R., Francaviglia, R., Aguilera, E., Smith, P.: Soil carbon sequestration rates under Mediterranean woody crops using recommended management practices: a meta-analysis. Agr. Ecosyst. Environ. **235**, 204–214 (2016)
18. Mayer, A., Hausfather, Z., Jones, A.D., Silver, W.L.: The potential of agricultural land management to contribute to lower global surface temperatures. Science Advances **4** (2018)
19. Alvaro-Fuentes, J., Morel, F.J., Plaza-Bonilla, D., Arrue, J.L., Cantero-Martinez, C.: Modelling tillage and nitrogen fertilization effects on soil organic carbon dynamics. Soil Tillage Res **120**, 32–39 (2012)
20. Francaviglia, R., Di Bene, C., Farina, R., Salvati, L.: Soil organic carbon sequestration and tillage systems in the Mediterranean Basin: a data mining approach. Nutr. Cycl. Agroecosyst. **107**, 125–137 (2017)
21. Almagro, M., Garcia-Franco, N., Martínez-Mena, M.: The potential of reducing tillage frequency and incorporating plant residues as a strategy for climate change mitigation in semiarid Mediterranean agroecosystems. Agr. Ecosyst. Environ. **246**, 210–220 (2017)
22. Alvaro-Fuentes, J., Paustian, K.: Potential soil carbon sequestration in a semiarid Mediterranean agroecosystem under climate change: quantifying management and climate effects. Plant Soil **338**, 261–272 (2011)
23. Agnelli, A., Bol, R., Trumbore, S.E., Dixon, L., Cocco, S., Corti, G.: Carbon and nitrogen in soil and vine roots in harrowed and grass-covered vineyards. Agric Ecosyst Environ **193** (2014)
24. Steenwerth, K., Belina, K.M.: Cover crops enhance soil organic matter, carbon dynamics and microbiological function in a vineyard agroecosystem. Appl Soil Ecol **40** (2008)
25. Alvaro-Fuentes, J., Lopez, M.V., Arrue, J.L., Moret, D., Paustian, K.: Tillage and cropping effects on soil organic carbon in Mediterranean semiarid agroecosystems: testing the century model. Agr. Ecosyst. Environ. **134**, 211–217 (2009)
26. Lugato, E., Bampa, F., Panagos, P., Montanarella, L., Jones, A.: Potential carbon sequestration of European arable soils estimated by modelling a comprehensive set of management practices. Glob. Change Biol. **20**, 3557–3567 (2014)
27. Soleimani, A., Hosseini, S.M., Bavani, A.R.M., Jafari, M., Francaviglia, R.: Simulating soil organic carbon stock as affected by land cover change and climate change, Hyrcanian forests (northern Iran). Sci. Total Environ. **599**, 1646–1657 (2017)
28. Lungarska, A., Chakir, R.: Climate-induced Land use change in france: impacts of agricultural adaptation and climate change mitigation. Ecol. Econ. **147**, 134–154 (2018)
29. Blanco-Canqui, H., Francis, C.A., Galusha, T.D.: Does organic farming accumulate carbon in deeper soil profiles in the long term? Geoderma **288**, 213–221 (2017)
30. Aguilera, E., Lassaletta, L., Gattinger, A., Gimeno, B.S.: Managing soil carbon for climate change mitigation and adaptation in Mediterranean cropping systems: a meta-analysis. Agr. Ecosyst. Environ. **168**, 25–36 (2013)
31. Brunori, E., Farina, R., Biasi, R.: Sustainable viticulture: the carbon-sink function of the vineyard agro-ecosystem. Agric Ecosyst Environ **223** (2016)

32. Gonzalez-Sanchez, E.J., Ordonez-Fernandez, R., Carbonell-Bojollo, R., Veroz-Gonzalez, O., Gil-Ribes, J.A.: Meta-analysis on atmospheric carbon capture in Spain through the use of conservation agriculture. Soil Tillage Res **122**, 52–60 (2012)
33. Yigini, Y., Panagos, P.: Assessment of soil organic carbon stocks under future climate and land cover changes in Europe. Sci. Total Environ. **557**, 838–850 (2016)
34. Department of Agriculture, Government of Catalonia: https://agricultura.gencat.cat/ca/serveis/cartografia-sig/aplicatius-tematics-geoinformacio/sigpac/
35. IPCC, Chapter 10: Emissions from livestock and manure management. https://www.ipcc-nggip.iges.or.jp/public/2006gl/pdf/4_Volume4/V4_10_Ch10_Livestock.pdf
36. Directive, N.: Council Directive 91/676/EEC of 12 December 1991 concerning the protection of waters against pollution caused by nitrates from agricultural sources. Official J. EUR-Lex **375**(31), 12 (1991)
37. Kamilaris, A., Engelbrecht, A., Pitsillides, A., Prenafeta-Boldu, F.X.: Transfer of manure as fertilizer from livestock farms to crop fields: the case of catalonia. Comput. Electron. Agri. J. **175**, (105550) (2020)
38. Kamilaris, A., Engelbrecht, A., Pitsillides, A., Prenafeta-Boldu, F.X.: Transfer of manure from livestock farms to crop fields as fertilizer using an ant inspired approach. In: XXIVth International Society for Photogrammetry and Remote Sensing (ISPRS) Congress, Nice, France (2020)
39. Funes, I., Savé, R., Rovira, P., Molowny-Horas, R., Alcañiz, J.M., Ascaso, E., Herms, I., Herrero, C., Boixadera, J., Vayreda, J.: Agricultural soil organic carbon stocks in the north-eastern Iberian Peninsula: Drivers and spatial variability. Sci. Total Environ. **668**, 283–294 (2019)
40. Institut Cartogràfic i Geològic de Catalunya (ICGC). https://www.icgc.cat/Administracio-i-empresa/Eines/Visualitzadors-Geoindex/Geoindex-Sols
41. Ubach, Teira: Avaluació i aprofitament dels residus orgànics d'origen ramader en agricultura. Quaderns de divulgació del LAF, núm. **5** (1999). https://ruralcat.gencat.cat/documents/20181/81510/Dejeccions2_49_Fertilit_organics_orig_ramader.pdf/8b5dea23-71b7-48de-bda6-c63596f0d440
42. Prenafeta-Boldú, et al.: Guia de les tecnologies de tractament de les dejeccions ramaderes a Catalunya. IRTA-DARP (2020)
43. Arco N, Romanyà J.: Guia de fonts de matèria orgànica apta per l'agricultura ecològica a Catalunya. Facultat de Farmàcia, Universitat de Barcelona (2010). https://pae.gencat.cat/web/.content/al_alimentacio/al01_pae/05_publicacions_material_referencia/arxius/guia_fonts_mo.pdf
44. Ministerio de Agricultura, Alimentación y Medio Ambiente, Gobierno de Espania, 2019. DECRETO 153/2019, de 3 de julio, de gestión de la fertilización del suelo y de las deyecciones ganaderas y de aprobación del programa de actuación en las zonas vulnerables en relación con la contaminación por nitratos procedentes de fuentes agrarias, https://dogc.gencat.cat/es/pdogc_canals_interns/pdogc_resultats_fitxa/index.html?action=fitxa&documentId=853461&language=ca_ES&newLang=es_ES.
45. Cherkassky, B.V., Goldberg, A.V., Radzik, T.: Shortest paths algorithms: theory and experimental evaluation. Math. Program. **73**(2), 129–174 (1996)
46. Lin, S., Kernighan, B.W.: An effective heuristic algorithm for the traveling-salesman problem. Oper. Res. **21**(2), 498–516 (1973)
47. Rotz, C.: Management to reduce nitrogen losses in animal production. J. Anim. Sci. **82**(13), E119–E137 (2004)
48. Fantappie, M., L'Abate, G., Costantini, E.A.C.: The influence of climate change on the soil organic carbon content in Italy from 1961 to 2008. Geomorphology **135**, 343–352 (2011)
49. Hoyle, F.C., O'Leary, R.A., Murphy, D.V.: Spatially governed climate factors dominate management in determining the quantity and distribution of soil organic carbon in dryland agricultural systems. Sci. Rep. **6**, 12 (2016)

50. Alcañiz, J.M., Boixadera, J., Felipó, M.T., Ortiz, J.O., Poch, R.M.: Chapter 12, Sòls. In: Martin Vide, J. (ed.) Tercer Informe sobre el Canvi Climàtic a Catalunya. Institut d'estudis Catalans i Generalitat de Catalunya, Barcelona, pp. 291–310 (2016)
51. Lawrence, C.R., Harden, J.W., Xu, X., Schulz, M.S., Trumbore, S.E.: Long-term controls on soil organic carbon with depth and time: a case study from the Cowlitz River Chronosequence WA USA. Geoderma **247**, 73–87 (2015)

Digital Twins, Augmented Reality and Explorer Maps Rising Attractiveness of Rural Regions for Outdoor Tourism

Peter Fischer-Stabel, Franziska Mai, Sabine Schindler, and Matthias Schneider

Abstract Hidden landscape features, visualized by very high resolution digital elevation models (DEM) in combination with augmented reality (AR) apps are able to rise the tourist attractiveness of hiking trails away from the known travel destinations, especially for outdoor adventure tourists. Within the work described, explorer maps as a new map format in combination with the enrichment of landscape perception using digital twins of geographic features were developed. One big challenge was the credible placement of the virtual objects in three-dimensional space. Natural markers at selected points of interest (POI) are used for the tracking of the device in the 3D room and the enrichment of the camera view with the hidden objects of the landscape (e.g. bomb craters, railway features), preprocessed before as virtual representation. This alternative view of the landscape has not only the potential to rise the attractiveness of destinations, but also to attract new target groups in hiking tourism, such as young people and technology-savvy adults.

Keywords Explorer map · Digital elevation model (DEM) · Outdoor adventure tourism · LiDAR

P. Fischer-Stabel (✉) · F. Mai
University of Applied Sciences Trier, Campusallee 25, 55761 Birkenfeld, Germany
e-mail: p.fischer-stabel@umwelt-campus.de

S. Schindler
Alea design, Hauptsr 55, 55767 Leisel, Germany

M. Schneider
Birkenfeld County, Schneewiesenstrasse 25, 55765 Birkenfeld, Germany

A. Kamilaris et al. (eds.), *Advances and New Trends in Environmental Informatics*, Progress in IS, https://doi.org/10.1007/978-3-030-61969-5_17

1 Introduction

1.1 Background

Current data from aircraft-borne laser scanning systems (ALS) are describing topographical objects and structures in natural and cultural landscapes in an unprecedented level of detail [1]. Therefore, LiDAR (light detection and ranging) makes it possible to break new ground in the tourist promotion of previously neglected travel regions as e.g. the Hunsrück region (SW-Germany) by highlighting hidden jewels in the field. An increase in tourist attractiveness especially for outdoor adventure tourists away from the known travel destinations can be achived.

Keeping this potential of LiDAR data in mind, the idea of "explorer" maps combining a very high resolution digital elevation model (DEM) and enriched with some topographic features, with a mobile augmented reality application (AR), was borne. On the one hand, the analogue and printed explorer map is showing micro-morphological structures at the earth´s surface (the landscape "naked"), inviting the visitor to detect hidden objects of natural, cultural or historical dimension (e.g. bomb craters, quarries, former settlements, mine shafts). On the other hand, the AR-App enriches the landscape experience with virtual objects displayed in the smartphone of the hiker.

A pilot project with the participation of the association of local history in the county of Birkenfeld, the institute for SoftwareSystems (University of Applied Sciences Trier), and the agency alea design, was initiated to develop a prototype of such discover maps. The so-called dream loop "Nohener-Naheschleife" along the internationally known long-distance hiking trail "Saar-Hunsrück-Steig" (SW-Germany) served as reference trail in order to pursue the basic idea formulated above and to lead it to practical usability.

1.2 Objective of the Project

The use of LiDAR data for outdoor adventure tourists will take place via two products realized:

- A new thematic map series, so-called "Explorer Map", which highlights the morphological features and microstructures in the region of the respective dream loop, as well as
- A complementary smartphone AR app that use augmented reality technology to visualize selected structures on site and provide background information to the hiker

Both products were implemented with the intention mainly to increase the attractiveness of the existing hiking destinations. In addition, new target groups for hiking tourism should be inspired to visit the region, especially the young it-affine generation. Details on trends regarding the hiking and adventure tourism can be found in [2].

Main objectives of the R&D project were:

- Development of a process for the appropriate cartographic visualization of micro-morphological landscape features from LiDAR data
- Design of a new type of map product "Explorer Map"
- Evaluation of the map product with regard to usability and attractiveness for outdoor vacationers
- Development of a workflow to develop AR - applications in landscape visualization using digital twins of geo-objects

The maps and apps are made available to interested visitors at suitable tourist information points on the respective reference routes. With help of a questionnaire, after their trail, the hikers are invited to evaluate the new designed maps and the accompaining apps regarding unsability and attractiveness. If the evaluations result is positive and the new analog and virtual map concept is accepted by the community, a statewide rollout is planned. The evaluation started with the first products in spring 2020.

2 Explorer Maps

The title "Explorer Map" already shows that it is a matter of a detailed disclosure—just discovery—of the peculiarities of the respective landscape hidden under the vegetation and overlayed in traditional maps.

Central idea behind the explorer maps is the visualisation of micro-morphological structures in the landscape based on a very high resolution digital elevation model (DEM). Figure 1 is showing an example: the aerial photo at the left side is showing in a traditional way the topographic information (e.g. forest, streets, water). The high resolution DEM at the right side visualizes the micromorphological structures below topography: e.g. two bomb craters in the center of the image, railroad embankments (lower right corner), alignments of country lanes (north part of the image).

While the processing of the high resolution DEM (horizontal ground resolution 20 × 20 cm, height accuracy around 20 cm) using the raw LiDAR point cloud data was done without difficulties, the finding of the illumination parameters to visualize the mirco-morphological structures of the landscape in an optimal way was challenging. Finally, we created two shaded reliefs with different illumination parameters. These two reliefs we blended for the final representation of the landscape model.

The final production of the analogue explorer map for hikers by adding the relevant supplementary topographic information and respecting the good practice rules in cartography [3] was done in a third step as a more or less routine operation.

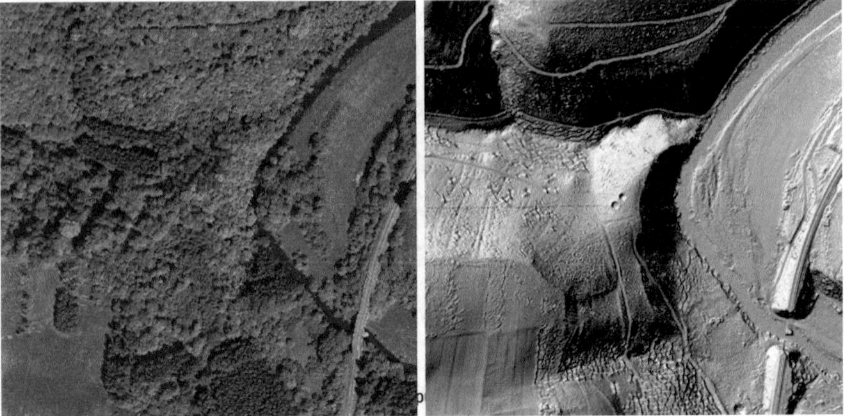

Fig. 1 Different representation of landscape surfaces by aerial photography and a shaded relief based on a high resolution digital elevation model

2.1 Reference Trail Nohener Naheschleife

As a first reference trail for the implementation, we choosed the hiking trail Nohener Naheschleife, one of the trails in Germany with the highest scores given by the German Hiking Institute. It represents a wonderful path through an almost untouched landscape that is only crossed by a railway line, looking like a "Märklin" landscape. On the hike, you are accompanied by impressive views into the wilderness of the Hunsrück low mountain ranges nature. The loop is also currently the only dream loop in Germany with a direct train connection [4].

Because each trail has it´s own pecularities, the further characterization in the title of the explorer maps encompasses the special topic. For example, the theme "Secret Railway, Rocks, River and Gorges" reveals the terrain surface of the dream loop Nohen with its natural but also the anthropogenic details (see Fig. 2). Based on this, the map picks up the theme-oriented, particularly characteristic terrain points and shows them enlarged in the map image. A legend based on the LiDAR map image explains further terrain structures.

3 Enrichment with Augmented Reality (AR)

Augmented Reality refers to a technology that makes it possible to expand the real environment of the user by inserting digital content. Conceptually, extensions for all human senses are conceivable, but in this project only the most important area, the visual AR, is dealt with. The type of possible extensions is very diverse: everything from text modules to complex 3D models is possible. Ideally, it appears to the user as if the real and digital objects exist in the same room. Even if the technology has

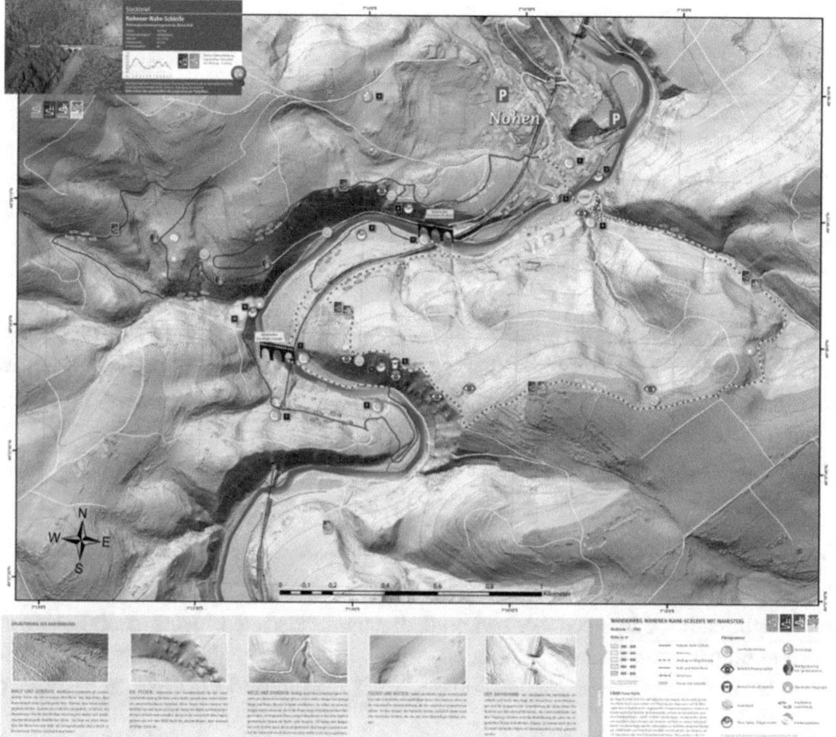

Fig. 2 Secret railway rocks river and gorges. Map window of the explorer map dream loop Nohener Naheschleife

only been accessible to a broader public in recent years, the basic concept has existed for a significantly longer time.

While the user immerses himself in a completely virtual world with Virtual Reality (VR), with Augmented Reality the real environment of the user remains present, e.g. through the image in the smartphone camera. However, this real existing environment is expanded by the integration of digital content.

AR ideas for tourism are as diverse as the tourism sector itself: individual city tours with context-related background information on history, architecture etc. are already offered as digital city guides, or digital visitor offers are provided to support the tasks and goals of a national park just to name a few applications. An overview regarding the use of digital twins and augmented reality in the touristic sector can be found in [5, 6].

While digital travel guides with AR support are now implemented in many urban destinations, there are currently hardly any practical solutions in the area of digital AR hiking guides [7]. The proposed project therefore intends to point out the possibilities of AR apps in connection with LiDAR data for outdoor activities on premium trails, and to close this gap.

Overall, some basic features can be identified in all augmented reality applications, that are also relevant in this project:

- Credible placement of the virtual objects in three-dimensional space
- Real time and interactivity

The type of presentation in which the digital content is made available to the user is particularly important for the construction of an AR system. For the project discussed here, it is the visualisation on a video display, in which the digital content becomes visible through the overlay on the live video (e.g. camera preview of a smartphone).

A common smartphone or tablet in combination with appropriate AR software is sufficient for this type of AR display. The user utilizes the device as a kind of a window into the landscape through which he views his surroundings. The digital content is displayed then in real time over the video image of the device.

3.1 Digital Twins of Landscape Objects

In the application described, we picked dedicated landscape objects as digital twins representing micromorphological peculiarities for the virtual enrichment. For this use case, the pre-processed LiDAR-based high resolution DEM had to be refined and further processed:

The regions of interest were clipped using the software QGIS and exported as *.TIF—File into Blender. There, the TIF file is loaded as a grid texture, resulting in a model whose height differences can be varied, texture be placed and edited. The finished model can be saved as a 3D model (*.fbx) or as a 2D view (*.jpg, *.png). When exporting as a 3D model, it is necessary first to "bake" the placed texture so that it is not lost during the export. In the 2D view, a camera must be placed in the model to get the desired view. After these processes, Blender is able to render the view, which can be saved as.png or.jpg—document (Fig. 3).

3.2 Tracking

To do the enrichment of the camera view, the position of the model must be constantly recalculated while the viewing angle changes and adjusted accordingly. This requires a real-time estimate of the position of the device in the room (tracking). The quality of the tracking is crucial for a credible positioning of the virtual objects [8].

Probably, the simplest approach to do the tracking is to use optical markers, also called fiducial markers. These markers are usually made up of black and white patterns, similar a normal QR code (Quick Response Code). With methods from computer vision, the camera image is first examined. If one of the previously defined markers has been found, the relative position of the camera can be determined based

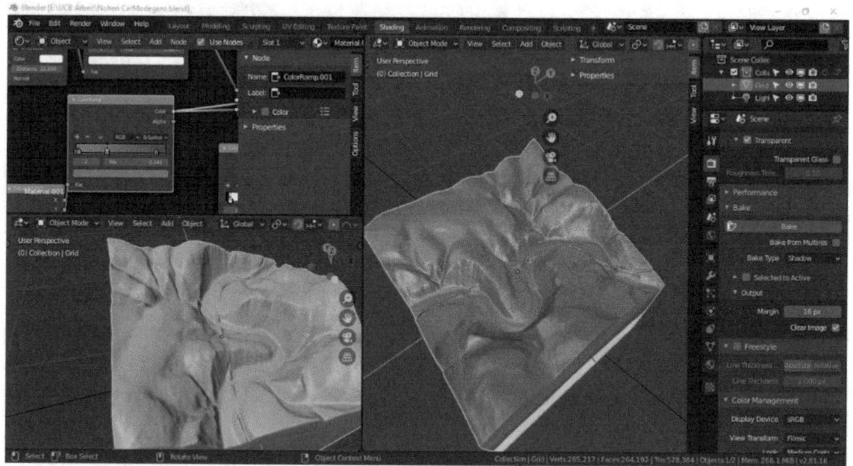

Fig. 3 Processing of the height map with the Blender software

on the position of the corner points. The calculated camera position then will be transferred to the digital camera.

In addition to the simple image markers, other images (natural markers) and textures are used for tracking. For this, specific features first must be extracted from the respective image. Typically, algorithms from the OpenCV framework, such as edge detectors, are used for this operation. If enough features are recognized in the camera image, the camera pose can be estimated in the same way as it is done by a marker-based tracking. It is therefore important that the selected image has a sufficient number of high-contrast feature points. Contrast poor images with repetitive structures are not suitable. One major advantage of feature-based tracking is the fact, that there are no separate marker necessary in the field.

In addition to the optical methods described above, the position of the mobile device can also be determined using internal sensors. Many mobile devices have inertial sensors (IMU), with which the angle of inclination of the device can be determined. In this regard, the so-called gyro sensor often is used in mobile devices. All three degrees of freedom can be measured using this sensor, but only relative to the original position. In combination with the acceleration sensor and magnetometer, the absolute orientation of the device in the global coordinate system can be determined.

Positioning sensors such as GPS (Global Positioning System) or GLONASS (Globalnaja nawigazionnaja sputnikowaja Sistema) are also able to determine the degrees of freedom of translation. Unfortunately, the positional information provided by the global navigation satellite systems (GNSS) sensors are not accurate enough to fulfil the requirement of a credible placement in our application.

Within the work described in the paper, we choosed natural markers (see Fig. 4 above) at selected points of interest (POI) to enrich the camera view with the hidden objects of the landscape (e.g. bomb craters, railway features). These objects processed

Fig. 4 Natural markers and target features

and prepared before in the Lab are positioned in the camera view of the smartphone as virtual representation at that selected POI (camera view direction natural marker).

These augmented or enriched views are highlighted in the printed version of the explorer map. Giving this information in the map, the hiker knows at what places along the trail there is a virtual enrichment of the camera view implemented and available in the field.

Since all the data and models needed are transferred to the mobile device the time of the download of the App, connectivity to a background server is not recommended in the field. This solution was preferred, because in the countryside, there is often a lack in the network connectivity but also the appropriate bandwidth.

3.3 App Development Environment

The technical requirements defined by some pilot users were relatively easy to fulfill: the software should work at an Android smartphone with at least Android 7.1 "Nougat". An integrated camera and a loudspeaker have to be available. In addition, the app should be able to be downloaded from the Google PlayStore (accessible via search bar or a QR code) (Fig. 5).

The app was developed under Windows for Android (API level from 25 upwards (API: Application Programming Interface). This corresponds to the development

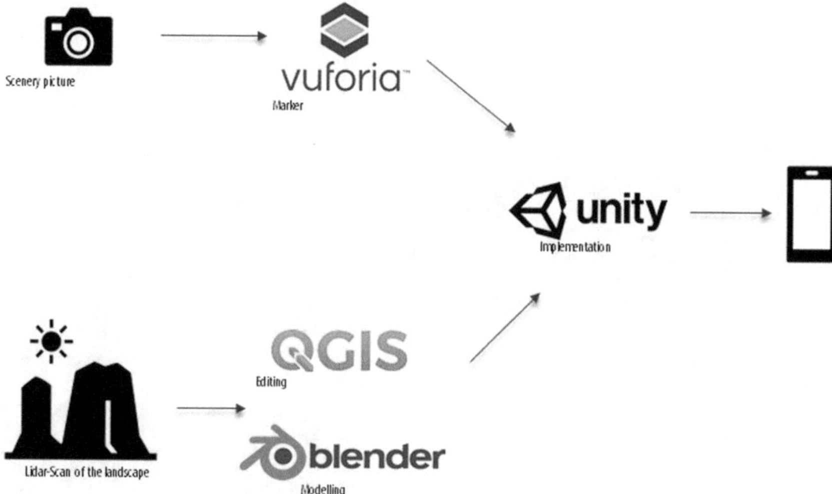

Fig. 5 App development environment

status of Android 7.1 Nougat from October 2016. The whole development environment was compounded by QGIS (production of the high resolution height maps), Blender (digital twins of landscape features), Vuforia as a software development kit (SDK) for augmented reality applications (creating and managing data and markers) and Unity. In addition, Photoshop was used to edit the 2D views because there is always a difference in perspective and distortion between model view and on-site photos.

As workflow, the markers are loaded from Vuforia as a database into Unity and used as an Image Target. This Image Target is assigned to the associated model or 2D view as a "child element". The 2D views often have to be further processed, distorted and color-edited in Photoshop so that their features are fitting well together. The 3D models require an additional light source and have to be scaled appropriately. In Unity, the marker and the model can be moved and rotated using the Image Target so that they fit well on top of each other.

Since the app is not officially available in the Google PlayStore, it is transferred to the desired Android device via activated USB debugging. In the demonstrator application developed, a 3D model is superimposed in a scene of the app, or, from a defined viewpoint overlayed with a 2D view as a superimposed layer in a separate scene (see Fig. 6).

Fig. 6 Screenshot of one AR-View of the railway banket

4 Discussion and Outlook

High-resolution terrain images derived from LiDAR point cloud data are able to give an alternative view of the earth´s surface. Special geographic features such as e.g. anthropogenic traces, but also natural objects, hidden under the vegetation coverage, are able to attract hikers to explore the landscape off the beaten tracks. This alternative view, preprocessed and manifested in the new explorer maps, and with the integration of new technologies such as augmented reality has the potential to rise the attractiveness of destinations for hiking tourists, but also to attract new target groups in hiking tourism, such as young people and technology-savvy adults.

From the authors perspective, the concept presented is able to give municipalities the opportunity to inexpensively enrich their hiking trail network and rise its attractiveness significantly in addition.

Unfortunately, the results of the survey in the hiker´s community regarding usability and attractiveness of the tools proposed are not yet available. If the evaluation will be positive, further thematic topics such as Celts, Romans, the Middle Ages, mining and geology, rocks and gems or coal-firing—to name just a few—will be processed within the scope of the project presented in this paper.

The entire thematic diversity, which is offered to the hiker with the 111 dream loops in the Hunsrück region, opens up the possibility of raising seemingly "unappealing" landscapes to a new level of attractiveness for tourists using the new explorer maps.

Reference

1. Wrozynski, R., Pyszny, K., Sojka, M.: Quantitative landscape assessment using LiDAR and rendered 360° panoramic images. Remote Sens. Open Access J. (2020). https://www.mdpi.com/2072-4292/12/3/386/htm. Last Accessed 06 Apr 2020
2. Godtman Kling, K., Fredman, P., Wall-Reinius, S.: Trails for tourism and outdoor recreation: A systematic literature review. Tourism Rev. **65**(4), 488–508 (2018)
3. Peterson, N.G.: GIS carography—a guide to effective map design. Taylor and Francis, New York (2009)
4. https://www.wanderinstitut.de/premiumwege/rheinland-pfalz/nohener-naheschleife/. Last Accessed 07 Apr 2020
5. D. Kim, S. Kim: The role of mobile technology in tourism: patents, articles, news, and mobile tour app reviews Sustainability **9**(11), 2082 (2017). https://doi.org/10.3390/su9112082
6. Yung, R., Khoo-Lattimore, C.: New realities: a systematic literature review on virtual reality and augmented reality in tourism research. Current Issues in Tourism **22**(17), 2056–2081 (2019). https://doi.org/10.1080/13683500.2017.1417359
7. https://www.viewranger.com/en-gb. Last Accessed 07 Apr 2020
8. Fischer-Stabel, P., Schneider, J., Göttert, C: Datenvisualisierung—Vom Diagramm zur Virtual Reality. utb-Verlag, Stuttgart (2018)

Identification of Tree Species in Japanese Forests Based on Aerial Photography and Deep Learning

Sarah Kentsch, Savvas Karatsiolis, Andreas Kamilaris, Luca Tomhave, and Maximo Larry Lopez Caceres

Abstract Natural forests are complex ecosystems whose tree species distribution and their ecosystem functions are still not well understood. Sustainable management of these forests is of high importance because of their significant role in climate regulation, biodiversity, soil erosion and disaster prevention among many other ecosystem services they provide. In Japan particularly, natural forests are mainly located in steep mountains, hence the use of aerial imagery in combination with computer vision are important modern tools that can be applied to forest research. Thus, this study constitutes a preliminary research in this field, aiming at classifying tree species in Japanese mixed forests using UAV images and deep learning in two different mixed forest types: a black pine (*Pinus thunbergii*)-black locust (*Robinia pseudoacacia*) and a larch (*Larix kaempferi*)-oak (*Quercus mongolica*) mixed forest. Our results indicate that it is possible to identify black locust trees with 62.6% True Positives (TP) and 98.1% True Negatives (TN), while lower precision was reached for larch trees (37.4% TP and 97.7% TN).

S. Kentsch · L. Tomhave · M. L. L. Caceres
United Graduate School of Agricultural Sciences (UGAS), Faculty of Agriculture, Yamagata University, Tsuruoka, Japan
e-mail: sarahkentsch@gmail.com

L. Tomhave
e-mail: luca.tomhave@kabelmail.de

M. L. L. Caceres
e-mail: larry@tds1.tr.yamagata-u.ac.jp

S. Karatsiolis (✉) · A. Kamilaris
Research Centre On Interactive Media, Smart Systems and Emerging Technologies (RISE), Nicosia, Cyprus
e-mail: karatsioliss@cytanet.com.cy

A. Kamilaris
e-mail: a.kamilaris@rise.org.cy

A. Kamilaris
Department of Computer Science, University of Twente, Enschede, The Netherlands

L. Tomhave
Leibniz University Hannover, Hannover, Germany

A. Kamilaris et al. (eds.), *Advances and New Trends in Environmental Informatics*, Progress in IS, https://doi.org/10.1007/978-3-030-61969-5_18

Keywords Aerial photography · Classification · Deep learning · Forestry · Mixed forests · Tree species

1 Introduction

Natural mixed forests are known as complex ecosystems with high resilience, high biodiversity, productivity and their carbon sink capacity. They play a role in the exchange of water carbon and nutrients within the soil-forest-atmosphere continuum. Under the present climate change conditions, high CO_2 emissions and degrading forest areas, made it essential to quantify the role of natural mixed forests on ameliorating the negative impact of anthropogenic emissions on climate change. Furthermore, the preservation of biodiversity, the physiological tolerances of species and effects of plant stress (due to droughts, pests and invasion) have to be considered. In particular, forests provide wood and non-wood resources, maintain soil fertility, regulate climate and preserve water supplies [1, 2]. Several studies proposed that a sound monitoring, stand inventories, quantification of tree species and ecosystem services are necessary to ensure their sustainability [3–5].

Forests in Japan occupy nearly 70% of the total territory. Two-thirds of the forests are located in mountainous areas and half of the total forest area made of timber plantations [6]. Forest plantations have a long history of clear-cuts followed by reforestation. Those planted forests with their simple structure of trees and lower biodiversity [7] have replaced natural forest areas, leading to a decrease in tree species diversity. Recent efforts to restore natural forests, as a result of climate change, have influenced Japan's point of view [8]. Recreation and protection are now the main drivers for forest management efforts [6]. However, natural forests have not fallen into adequate management strategies [8]. Furthermore, natural mixed forests have not been adequately studied, while most of the existing relevant case studies were only carried out for small forest patches [9]. Thus, tree species' composition and diversity as well as their distribution and interaction within the forest ecosystem need to be studied. The first step is to develop a monitoring system that allows a rapid and reliable method to survey this type of forest and additionally to be able to identify tree species. The use of aerial imagery in combination with deep learning approaches are essential tools and techniques that can build the bases for the improvement of monitoring methodologies in forestry research [10–13]. We used UAV (Unmanned Aerial Vehicles) to capture images of the forests and then trained a deep learning network to identify tree species in two different forest types. Specifically, we used one dataset to identify invasive black locust trees in coastal black pine forests from drone images and a second dataset to identify conifer trees. Since this work is the first attempt of our research team to combine forestry and AI, the main purpose is to evaluate the results of automatic tree species' identification.

2 Related Work

Previous studies used aerial images gathered by satellites, airplanes and UAVs to identify trees species in forests. The work in [14] classified tree species in a mixed forest by using high-resolution IKONOS data. Twenty-one species were classified by using panchromatic and multi-spectral bands. After pre-processing the images, 50 pixels per species were extracted and a Turkey´s multiple comparison test was applied. Finally, maximum likelihood classifiers were performed for reaching accuracies of 62%. According to [14] broadleaf trees are more difficult to classify than conifers.

In a comparison study of tree identification using IKONOS and WorldView-2 (WV2) images with a resolution of 1–4 m, overall accuracies of 57% were reported for 7 tree species and 15 selected features [15]. In this study, linear discriminant analysis (DLA) and decision tree classifier (CART) were used.

Dalponte et al. [16] used both high-resolution airborne hyperspectral images and satellite images, each in combination with LIDAR data to understand the classification potential by using different datasets. The data was pre-processed for normalization and generalization, as well as feature selection for LIDAR data. Support Vector Machines (SVM) and random forest classifiers (RFC) were used for the classifications. Different classes were tested, ranging from single tree species to macro-classes reaching kappa accuracies of 76.5–93.2%, concluding that hyperspectral data resulted in highest accuracies while SVM outperformed RFC. Torresan et al. [17] reported that 14% of UAV-related studies focused on tree species classification. One such study was carried out in [18], classifying tree species by using RGB and hyperspectral images of a boreal forest. In total, 11 orthomosaics and DSMs (Digital Surface Models) were used, as well as reference data. Spectral and 3D point cloud information was used to classify trees. RFC and multilayer perceptrons performed the classification with accuracies of around 95% for four different tree species. Moreover, the work in [19] focused on the best time-window to gather images by UAVs. Their primary aim was to effectively classify tree species using a multi-temporal dataset. The data was gathered in a broadleaf forest, composed by 577 tree species that were divided into 5 groups. Orthomosaics were used to analyze the spectral response, the characteristics and differences of tree classes. The pixel intensity was used to run RFC. Misclassifications of 15.9 and 36% were observed by using one dataset of one season only. The error was decreased to 8.8% when using multitemporal datasets. A method to classify tree species in a mixed forest dominated by pine trees using high-resolution RGB images collected by three-years was proposed in [20]. Orthomosaics and DSMs were used to delineate tree crowns in a first step by using local maxima filtering the watershed algorithm. The extracted tree crowns were used to train a Convolutional Neural Network (CNN). Therefore, two approaches were conducted: one using one orthomosaic and another using three orthomosaics for training and/or testing. Classification accuracies between 51–80% were recorded.

3 Methodology

3.1 Problem Description

The first study site, where the first dataset was created, is located in the Yamagata prefecture on the Japanese northwest coast (38°49′14"N, 139°47′47"E). The coastal forest is a black pine (*Pinus thunbergii*) plantation (Fig. 2a) with high tolerances against acidity, alkalinity and salty soils and drought conditions. This forest was planted in order to protect the surrounding area from strong winds and sand movement. Since the early 1990′s, this forest has been invaded by black locust trees (*Robinia pseudoacacia*) (Fig. 2b), a fast-growing species establishing in gaps of the black pine forest. Black locust species is known for its rapid invasion and high biomass production with a high impact on the structure and function of tree communities [21]. However, the exact influences on the functions of the coastal forest are unknown. Generally, invasive tree species have a high impact on the structures, properties and functions of natural ecosystems [22, 23]. Therefore, it is necessary as a first step to detect and identify black locust trees in order to provide information about their distribution and density, as a second step, to understand the structure and nutritional impact of this invasion on the black pine properties as a windbreak and growth. Information about these parameters will offer essential insights for a sound management of this type of forest. Additionally, this information offers the possibility to quantify the effects of invasive species as it turns the monoculture into a mixed forest, which is supposed to be less affected by diseases and infections [24]. The second site, which provides the second dataset, is located in the Yamagata University Research Forest (YURF) on the Japanese main island Honshu. It is located in the northern part of the Asahi Mountains. The research forest covers an area of 753 ha.

The forest is characterized by steep slopes (30–44 degrees) within a range of altitudes between 250 and 850 m and it is crossed almost in half by the Wasada River. The area is composed of a mixed natural forest, as well as deciduous broadleaf and coniferous trees [25]. Our study sites are located in a slope, as shown in Fig. 1. The second site is mainly composed of a mixture of larch (*Larix kaempferi*) trees and oak (*Quercus mongolica*) and in a minor proportion beech (*Fagus crenata*) (Fig. 2). Such as the YURF, the large majority of Japanese forests is located in steep mountains and is characteristic by a complex ecology. Thus, the conduction of field surveys is very much limited, primarily due to the lack of accessibility and the necessity of man-power. Therefore, developing an automatic methodology capable of identifying tree species in mixed forests is crucial for the evaluation of these forest ecosystems.

3.2 Data Collection

Data acquisition was carried out using a DJI Phantom 4 drone. The drone is equipped with a 12-megapixel camera which produces high resolution geo-referenced images.

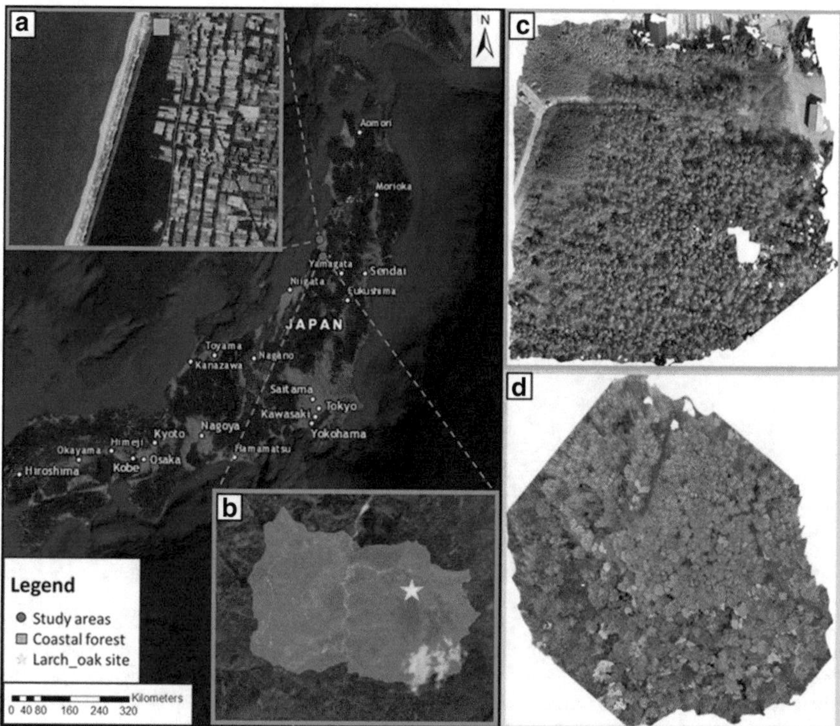

Fig. 1 The main map showing the location of the study areas in Japan, where data was acquired: **a** Coastal forest near Sakata city. **b** YURF (orange area), located south of Tsuruoka city in the Asahi mountains. **c** Orthomosaic of the coastal forest, where images were collected to identify black locust trees that are invasive in this area. **d** Orthomosaic of the larch oak site, where the minor tree species larch should be identified

The flights were performed using the autonomous mode, which standardizes the acquisition protocol. The coastal forest was photographed in July and October 2019, but only the analysis of images from October are presented in this study. Images from July failed to be automatically recognized (i.e. in a previous analysis we performed) since differences between tree species were too low. During the flight, around 1100 images were collected with an overlap of 90%. The flight altitude was 30 m and the covered area was 2.7 ha. Since we already faced difficulties in identifying trees in the first case study, we chose from the beginning autumn images for the second one. Manual annotations were only available for this season since the forest expert experienced the same limitations in accurately identifying tree species in the summer orthomosaics. For the larch-oak site, an additional flight was performed at the end of October 2019, capturing 202 images with a set overlap of 93% covering an area of 3.2 ha. The collected images were processed with Agisoft Metashape [26], in order to generate the orthomosaics. The autumn images of the coastal forest were not aligned well and the orthomosaics had several blank spots. The resolution of the

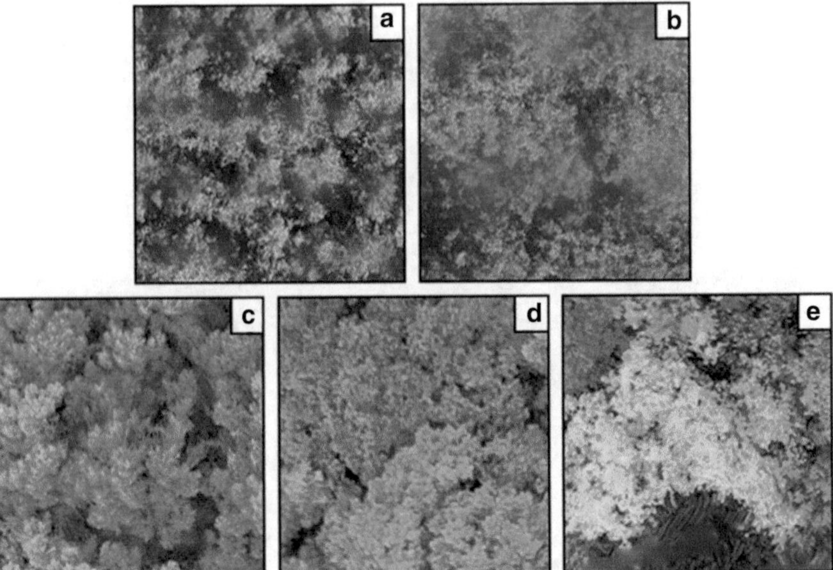

Fig. 2 Examples of tree species in our study areas, as shown from aerial photos. **a** Pine trees and **b** black locust represent trees of the coastal forest. The trees look similar to each other with small differences in the colour (due to the colour change of black locust trees), which increases the difficulty to identify them. **c** Larch, **d** oak and **e** beech represent trees of the YURF mixed forest. As can be seen in the images, autumn images were chosen since the trees show different coloured leaves which increase the potential to classify them

orthomosaics was approximately 1.1 cm/pix for the coastal forest sites and 3.5 cm/pix for the larch-oak sites.

3.3 Data Pre-processing

The two datasets were annotated by forest experts, knowing the study area well. The annotations were done using the image editing software Gimp. Areas of each tree species were colored black and stored in different layers. Those layers were used as ground truth data. Due to the different visual characteristic of larch trees, the accuracy of that layer was assumed as high. More difficulties appeared during the annotation process of the coastal forest site due to the comparably small canopy area of the black locust and its mix with other broadleaved species. Furthermore, since a new approach for tree classification was used by the experts, a certain misclassification cannot be ruled out. Nonetheless, a high accuracy of annotations can be assumed for both sites.

Image segmentation aims to partition an image into semantically related segments so that each pixel in the image is assigned to a group of coherent pixels. Consequently, this enables image analysis focusing on its elements. The specific tree identification problem differs from general-case image segmentation in the sense that each image pixel represents a larger physical space ($\approx 1m^2$) and does not necessarily belong to a single component of a larger entity. Roughly, each pixel in a species map corresponds to the foliage of a single tree. However, trees are more likely to grow next to trees of the same species, forming groups that cover significant ground surface represented by many pixels in an aerial photo of the investigated area.

The input of the model accepts patches of size 64 × 64 so the image of some forest under examination is partitioned into patches of the specific size. Accordingly, the accompanying tree species' map is partitioned to patches of the same size that correspond to the same forest area. Furthermore, the tree species' maps are converted to binary maps with pixel values of zero or one that indicate whether the specific pixel corresponds to the target species or not. This approach creates a dataset comprised from several thousand patches and their corresponding maps. We applied no special pre-proccessing to the input patches but we excluded patches that contain very few or no trees at all. More specifically, we detected little or no representation of trees by examining the total brightness of each patch since ground texture tends to be much brighter than foliage. An appropriate threshold for deciding which patches to discard because of reduced trees depiction was determined after visual inspection.

3.4 Deep Learning Modelling

Both image segmentation and tree identification make use of an input image and an accompanying pixel-wise map that holds the labels of the represented trees. For image segmentation, labels correspond to image objects and subjects while, for tree identification, labels correspond to tree species.

Model selection: The most popular state-of-the-art models for image segmentation are the fully convolutional Dense-Net [27], the multi-scale context aggregation by dilated convolutions [28], the DeepLab model that uses spatial pyramid pooling [29], the FastNet [30] and the U-Net model [31]. The latter was used in biomedical image segmentation and yielded precise results in tasks where getting a class label for each pixel is crucial. Despite its effectiveness, U-Net is very straight-forward to implement and does not require extreme fine-tuning and task-specific architectural modifications. Given the fact that we deal with tasks comprising from binary labels (since we try to distinguish between two tree species at a time), U-Net architecture is considered sufficient for tackling the problem. The name of the U-Net model comes from its shape which is formed by its two data flow paths: the contrastive and the expansive paths as shown in Fig. 3. Input goes through the contrastive path consisting of subsequent convolutional and down-sampling operations until it is considerably reduced in size at the center of the model. After that point, feature maps are processed by subsequent convolutional and up-sampling operations until the size of the labeled maps is reached.

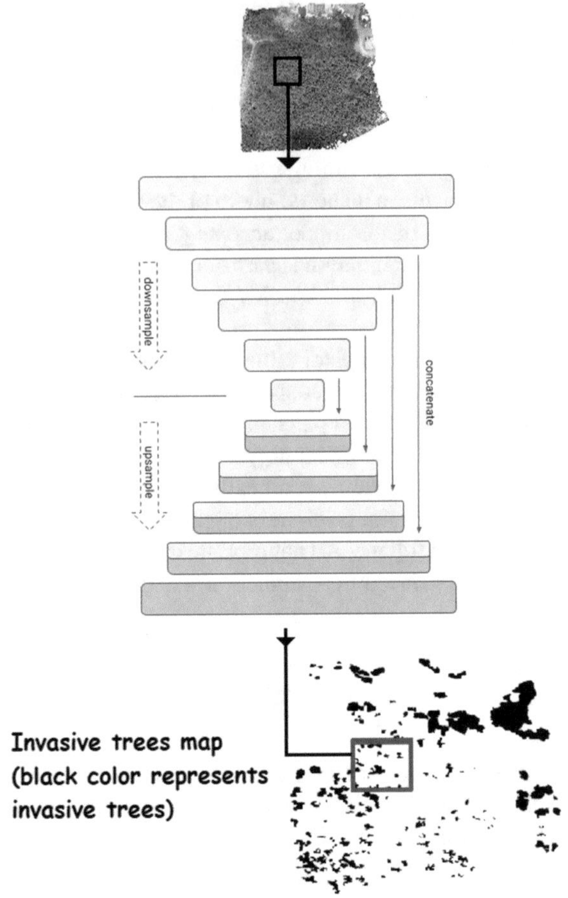

Fig. 3 The invasive tree identification model with a U-Net architecture. Patches of the orthomosaic were inputted in the model reduced in the size while passing the contrastive path and outputted after resizing and labelling as invasive trees map

Design decisions: The predicted output and the actual output (labeled map) contribute to the loss function of the model, which is cross-entropy. All model layers use the ReLU activation function [32] except the final output layer that uses the sigmoid function, which serves the requirement that every pixel in the output map has a probability of belonging to a certain label that is independent from any other pixel output.

Avoiding overfitting: An important architectural characteristic we used to avoid overfitting, caused by the relatively small dataset and the high capacity of the model (consisting of *9,075,201* trainable parameters), is the two dropout [33] layers in the model's contrastive path. These layers randomly drop some elements of the feature maps and prevent co-adaptation of the parameters during training, enhancing the generalization of the model and avoiding overfitting. The model is trained on mini-batches of image patches as input and their maps of tree species identification as targets.

Dataset imbalance: Invasive tree species are much less in number than the dominant species of a forest. In the studied cases, invasive trees occupy less than 10% of the forests' surfaces. This dataset imbalance greatly reflects on the model performance if no countermeasures are applied: the majority class overwelmes the minority class (the invasive species) and the identification of the latter is extremely poor. To this direction, the loss function is weighted appropriately so that the minority class misclassification is penalized to the extend that the dataset imbalance effect cancels out. The loss function is weighted analogous to the number of the invasive trees in a patch: more invasive trees present in a patch translates to a higher misclassification penalty. The calculation for the loss weighting factor f_M of a invasive tree map M is shown below.

$$f_M = \frac{\sum_{i,j} M[i,j]}{a} + 1, \quad M[i,j] = \begin{cases} 1 \; if \; M[i,j] \, represents \; an \; invasive \; tree \\ 0 \; otherwise \end{cases}$$

The value of a is determined using the ratio of the total number of invasive trees in the dataset over the total number of trees not considered invasive. This ratio is multiplied by the number of pixels in each tree map ($64 \times 64 = 4096$) to provide the value of a. Since this ratio is about 10% in the available datasets, we use $a = 400$.

Data augmentation: Image augmentation is also applied to enhance the performance of the model. Since the labels for the invasive tree identification problems are provided in a 2-D species map, any spatial transformation applied to the input patches must also be applied to the target map. For example, an image rotation of the input patch by definition requires an image rotation of the species map in order to maintain the one-to-one correspondence between the input image pixels and the pixels of the output map. To avoid practical problems due to interpolation schemes on the species maps, like distorting their binary values, we only use spatial transformations that do not require pixel interpolation. Such augmentations are image rotations by angles that are multiples of *90* degrees and vertical or horizontal image flipping. We also use four other augmentation schemes based on non-spatial transformations: image sharpness, brightness and color adjustments and random color channel shifting by a small amount.

Performance metrics: The model performance is measured by simple statistics like false and true positive/negative classification ratios. We particularly measure the percentage of output map pixels that are classified as an invasive species while actually being one (True positive) and the percentage of output map pixels that are classified as invasive species but actually are not (False Positive). These metrics are more appropriate than plain classification success rate because of the class imbalance. The decision threshold of the output neurons is also shifted upwards in order to obtain a good compromise among the performance metrics.

Data split for testing: Tuning of the output threshold was performed using a validation set. This essentially means that the dataset is split into a training set comprised from 60% of the available data, a validation set comprised from 20% of the data and a test set containing the remaining data. During the creation of the various sets, special care was taken to make sure that all sets contain a similar number of samples in terms of invasive species surface coverage. Specifically, the invasive tree percentage coverage of the tree species maps was noted for every patch in the dataset and was categorized as 0% invasive tree coverage, invasive tree coverage between 1–20%, 21–50%, 51–80% and invasive tree coverage in the range 81–100%. During the split of the patches to the three sets, an equal ratio of the coverage categories was represented in each set. So, the training set, validation set and test set hold an equal percentage of the invasive tree coverage categories. We found that following this balancing approach instead of an n-fold validation approach significantly improved model generalization. The model is trained with the stochastic gradient descent algorithm with a learning rate of 1×10^{-3}.

4 Results

We examined two cases of tree species identification. The first deals with detection of black locust (site 1 as described in Sect. 3.1); the second with detection of larch trees (site 2, described in Sect. 3.1). The validation process determined 0.85 to be an appropriate output threshold value for both cases. Results are shown in Table 1.

Larch identification has a significantly lower success rate because the corresponding dataset is much smaller than the dataset showing black locust trees. The results are greatly affected by the threshold value of the output layer. Raising the threshold value reduces the correct classification rate of the target tree species and improves the correct classification rate of the non-target trees. For a better model evaluation, we also examine the effect of the invasive trees' proportion within a specific image patch on the classification of at least the number of invasive trees that corresponds to the imbalance ratio between the tree species ($\approx$ 10%). In other words, we observe the rate at which the model identifies at least 10% of the invasive trees in a patch, in relation to the percentage of the visible invasive trees in the patch. The black locust test set patches are divided into five categories according to the surface covered by the containing invasive trees: zero invasive trees; less than 20% of the total pixels of the 64×64 patch; invasive trees covering 20–49% of the total surface; 50–79% surface coverage; and 80–100% coverage. For each patch that the model identifies at least 10% of invasive trees contained in it, we consider it as being classified correctly. Table 2 shows the results of this experiment.

Table 1 Results of the U-Net model on the black locust and the larch identification problem. True positive and True negative values represent the quality of the results of the two identification case studies, black locust and larch trees

	Black locust (%)	Larch (%)
True positive	62.6	37.4
True negative	98.1	97.7

Table 2 Identification percentage of at least 10% of the contained invasive trees in the test set patches in relevance to the percentage of pixels occupied by invasive trees

Invasive trees coverage (%)	0	1–19	20–49	50–79	80–100
Detection of at least 10% of contained invasive trees (%)	97.04	18.88	27.54	48.8	61.7

As expected, the blacker locust trees in a patch, the better chance for the model to detect at least the portion of them that corresponds to their statistical frequency of occurrence. Invasive tree clusters triggered the detection of appropriate features more often than scattered invasive trees.

5 Discussion

The methodology was successfully applied to our datasets, even though there is much space for improvements. We acknowledge the low detection rates, which are mainly related to the insufficient amount and balance of the training data. The results of the first dataset indicated that the amount of data for black locusts is not enough to train a deep learning network. Moreover, the data are imbalanced regarding the classes black pine and black locust, which can be improved by increasing the dataset. In Japan, black locust trees appear mixed with other tree species which makes it difficult to get images from pure black locust stands. Nevertheless, since the invasion of the tree species is a problem in forests all over Japan (e.g. [34, 35]) images of black locust in mixed forests can be easily collected. Our second dataset dealt with images of the mixed forest in YURF. We ran the deep learning model for only one of the orthomosaics to get a first idea of the precision we could achieve. We assume that the accuracy of the deep learning application can be increased by using data from the whole season (spring, summer and autumn).

5.1 Study Implications

Coastal forests in Japan have peculiar stand conditions (sandy soils, high salinities and strong winds) and both black pine and black locust tree species are adapted to these conditions. The invasion had changed the monoculture plantation into a mixed forest, which might lead to an enhancement of forest resilience, since several studies have pointed out the benefits of mixed forests under climate change. Tinya et al. [36] suggested the increase of stability in forest stands to stress and disturbances that can be mitigated by mixed forests. The image analysis performed offers the opportunity to study the benefits of the forest mixture and helps to characterize the resilience of this mixed forest. Even though the black locust can have positive effects on the coastal forest, negative influences are discussed as well. Since black locust trees are deciduous, they will provide wind tunnels in the leaf absent season. We assume that the distribution of the trees has a significant influence on the windbreak potential of the coastal forests. On the other hand, larch trees, which were studied via our second dataset, are part of a different kind of mixed forest. The focus of previous studies has been to evaluate processes in forests in relation to the tree species [37] or the composition of the forest stands [37, 38]. These studies tried to solve important aspects of forests and propose solutions for mitigating the effects of climate change, focusing though only on small forest patches. Thus, a reliable methodology for scaling up to forest stands can provide insights not only about tree species composition but also contribute to understand other essential forest characteristics (soil type, soil moisture, fungi, nutrient cycles, etc.). The classification of the mixed forest via computer vision, such as the work in this study, is important to further achieve these goals.

5.2 Policy-Making

The methodology proposed in this paper is a tool for forest management practices since it can provide fast and reliable information about forests. In particular, the coastal forests with their functions as wind and tsunami breakers could benefit from a more automated management system. Further, the Ministry of Environment in Japan has called for the urgency of management issues of black locust species [35]. Since most of the forests are unmanaged and dense, our study provides a simple tool for forestry to assess the spread of black locusts and provides the possibility to detect invasive trees. Furthermore, since mountainous forests are steep and hard to access, the methodology of this study partially solves the problem of inaccessibility. Since the technique uses merely images, this methodology can be applied for forestry classification/management world-wide, contributing to more focalized field surveys.

5.3 Limitations and Assumptions

Our study shows that we are able to identify tree species, however, we also face some limitations. In the first case study, the orthomosaic used showed less than 10% of black locust trees. Fieldwork in this area showed that the amount of black locust is higher than the 10%, as shown in the orthomosaic, which can be explained by their smaller height (12–18 m) in comparison to black pine trees (up to 40 m). The smaller black locusts are often covered by the black pine canopies but they are partly still visible since black locust trees mainly appear in gaps formed in between the black pines. The true distribution and number of trees is still unknown, since not all trees are visible on the images. Therefore, further fieldwork needs to be conducted to evaluate the maps generated by our model. Even though the method has its limitations it provides a fast overview of the study area and facilitates management approaches. In the second case, we attempted to identify larch trees in a mixed forest, where they are one of the dominant tree species. Larch tree structure makes it easier to recognize them from the images. The general idea of this approach was to see how well the deep learning network can deal with these images since a further step will be the classification of all dominant tree species in the mixed forest. Field surveys indicated that in our study areas there are several trees which are covered by canopies of taller trees. Thus, we acknowledge that this methodology might not get all the trees in the forest and may only provide information of the visible and dominant tree species. Since in situ fieldwork is barely possible in these areas and knowledge gaps about mixed forests are still large [24], our methodology can be considered as a helpful tool for forestry research.

5.4 Future Work

For future work, we aim to overcome some of the limitations mentioned by using satellite data to increase the amount of data used as input to our models. We plan to locate regions in North America where black locust is a native species and acquire satellite photos to augment our dataset and reduce the existing imbalance of the data. We also plan to acquire images of different seasons (spring, summer and autumn), which will help to increase the accuracy of the model, something already shown as a basic solution in related work [39]. In this context, the work in [40] demonstrated the effectiveness of a multi-temporal dataset on image classification issues. Generating synthetic data is another option worth considering [41, 42].

A comparison of satellite and drone images for tree species' classification for deep learning applications is an interesting aspect we also plan to work on, primarily for the black locust problem. Prior fieldwork results in the larch-oak mixed forest site showed that our study area includes more than 20 different tree species, although some of them are in small numbers. Therefore, we will focus on the dominant tree species of the mixed forest, namely by increasing the training data for larch, oak and beech trees.

Further approaches, for instance the additional use of spatial information along with the images are not considered, since natural forest structures, tree species' compositions and the behavior of the invasive tree species are irregular and not well-known.

6 Conclusion

Our study aimed to identify two kinds of tree species in Japanese mixed forest by using UAV-acquired RGB images and deep learning technologies. Our results indicated that it is possible to classify black locust and larch trees by using these two technologies. The model was able to identify patches without black locust/larch with high accuracies but showed lower accuracies for detecting the target species. The main reason was the imbalance in our input data. The number of images representing black locust and larch trees was significantly lower than the other trees. Even though our data were highly imbalanced, the results are promising for future work in this field, since our proposed methodology is suitable for large-scale forestry management applications. Further data acquisition efforts are planned for increasing our datasets and improve the performance of our model.

Acknowledgements Andreas Kamilaris and Savvas Karatsiolis have received funding from the European Union's Horizon 2020 research and innovation programme under grant agreement No 739578 complemented by the Government of the Republic of Cyprus through the Directorate General for European Programmes, Coordination and Development.

References

1. Thompson, I., Mackey, B., McNulty, S., Mosseler, A.: Forest resilience, biodiversity, and climate change: a synthesis of the biodiversity/resilience/stability relationship in forest ecosystems. Technical Series no. 43. pp. 1–67. Secretariat of the Convention on Biological Diversity, Montreal (2009)
2. Núñez, D., Nahuelhual, L., Oyarzún, C.: Forests and water: the value of native temperate forests in supplying water for human consumption. Ecolog. Econ. **58**(3), 606–616 (2006) ISSN 0921-8009. https://doi.org/10.1016/j.ecolecon.2005.08.010
3. Norton, D., Hamish Cochrane, C., Reay, S.: Crown-stem dimension relationships in two New Zealand native forests. New Zealand J. Bot. **43**(3), 673–678 (2005). https://doi.org/10.1080/0028825X.2005.9512984
4. Lara, A., Little, C., Urrutia, R., McPhee, J., Álvarez-Garretón, C., Oyarzún, C., Soto, D., Donoso, P., Nahuelhual, L., Pino, M., Arismendi, I.: Assessment of ecosystem services as an opportunity for the conservation and management of native forests in Chile. Forest Ecol. Manage. **258**(4), 415–424 (2009) ISSN 0378-1127. https://doi.org/10.1016/j.foreco.2009.01.004
5. Nahuelhual, L., Donoso, P., Lara, A., et al.: Valuing ecosystem services of chilean temperate rainforests. Environ Dev Sustain **9**, 481–499 (2007). https://doi.org/10.1007/s10668-006-9033-8
6. Knight, J.: From Timber to Tourism: Recommoditizing the Japanese Forest. Develop. Change. **31**, 341–359 (2000). https://doi.org/10.1111/1467-7660.00157
7. Ito, S., Nakayama, R., Buckley, G.P.: Effects of previous land-use on plant species diversity in semi-natural and plantation forests in a warm-temperate region in southeastern Kyushu. Japan. Forest Ecol. Manage. **196**(2–3), 213–225 (2004) ISSN 0378-1127. https://doi.org/10.1016/j.foreco.2004.02.050
8. Yamaura, Y., Oka, H., Taki, H., et al.: Sustainable management of planted landscapes: lessons from Japan. Biodivers Conserv **21**, 3107–3129 (2012)
9. Suzuki, W., Osumi, K., Masaki, T., Takahashi, K., Daimaru, H., Hoshizaki, K.: Disturbance regimes and community structures of a riparian and an adjacent terrace stand in the Kanumazawa Riparian research forest, northern Japan. Forest Ecol. Manage. **157**(1–3), 285–301 (2002) ISSN 0378-1127. https://doi.org/10.1016/S0378-1127(00)00667-8
10. Tang, L., Shao, G.: Drone remote sensing for forestry research and practices. J For. Res. **26**(4), 791–797 (2015) ISSN 1993-0607. https://doi.org/10.1007/s11676-015-0088-y
11. Paneque-Gálvez, J., McCall, M.K., Napoletano, B.M., Wich, S.A., Koh, L.P.: Small Drones for Community-Based Forest Monitoring: An Assessment of Their Feasibility and Potential in

Tropical Areas. J. For. **5**(6), 1481–1507 (2014) ISSN 1999-4907. https://doi.org/10.3390/f5061481

12. Gambella, F., Sistu, L., Piccirilli, D., Corposanto, S., Caria, M., Arcangeletti, E., Proto, A.R., Chessa, G., Pazzona, A.: Forest and UAV: a bibliometric review. J. Contemp. Eng. Sci **9**, 1359–1370 (2016)
13. Fromm, M., Schubert, M., Castilla, G., Linke, J., McDermid, G.: Automated detection of conifer seedlings in drone imagery using convolutional neural networks. Remote Sens. 11(21) (2019) ISSN 2072–4292. https://doi.org/10.3390/rs11212585
14. Katoh, M.: Classifying tree species in a northern mixed forest using high-resolution IKONOS data. J for Res **9**, 7–14 (2004). https://doi.org/10.1007/s10310-003-0045-z
15. Pu, R., Landry, S.: A comparative analysis of high spatial resolution IKONOS and WorldView-2 imagery for mapping urban tree species. Remote Sens. Environ. **124**, 516–533 (2012) (ISSN 0034-4257)
16. Dalponte, M., Bruzzone, L., Gianelle, D.: Tree species classification in the Southern Alps based on the fusion of very high geometrical resolution multispectral/hyperspectral images and LiDAR data. Remote Sens. Environ. **123**, 258–270 (2012) ISSN 0034-4257. https://doi.org/10.1016/j.rse.2012.03.013
17. Torresan, Ch., Berton, A., Carotenuto, F., Filippo, S., Gennaro, S.F., Gioli, B., Matese, A., et al.: Forestry applications of UAVs in Europe: a review. Int. J. Remote Sens. **38**(8–10), 2427–2447 (2016)
18. Nevalainen, O., Honkavaara, E., Tuominen, S., Viljanen, N., Hakala, T., Yu, X., Hyyppä, J., Saari, H., Pölönen, I., Imai, N.N., Tommaselli, A.M.G.: Individual tree detection and classification with UAV-based photogrammetric point clouds and hyperspectral imaging. Remote Sens. **9**, 185 (2017)
19. Lisein J, Michez A, Claessens H, Lejeune P.: Discrimination of deciduous tree species from time series of unmanned aerial system imagery. PLoS One **10**(11):e0141006 (2015).https://doi.org/10.1371/journal.pone.0141006
20. Natesan, S., Armenakis, C., Vepakomma, U.: Resnet-based tree species classification using UAV images. Int. Arch. Photogramm. Remote Sens. Spatial Inf. Sci. XLII-2/W13, 475–481 (2019). https://doi.org/10.5194/isprs-archives-XLII-2-W13-475-2019
21. Lopez C, M. L., Mizota, C., Nobori, Y., Sasaki, T., Yamanaka, T.: Temporal changes in nitrogen acquisition of Japanese black pine (Pinus thunbergii) associated with black locust (Robinia pseudoacacia). J. For. Res. 25(3), 585–589 (2014) ISSN 1993–0607. https://doi.org/10.1007/s11676-014-0498-2
22. Richardson, D., Binggeli, P., Schroth, G.: Invasive agroforestry trees–problems and solutions. Agroforestry and biodiversity conservation in tropical landscapes, pp. 371–396. Island Press, Washington, (2004)
23. Moran, V.C., Hoffmann, J.H., Donnelly, D., Wilgen, B.W. van, Zimmermann, H.G.: Biological control of alien, invasive pine trees species in South Africa. In: Proceedings of the X International Symposium on Biological Control of Weeds, Spencer, N. R., pp. 941–953. Bozeman, USA (2000)
24. Coll, L., Ameztegui, A., Collet, C., Löf, M., Mason, B., Pach, M., Verheyen, K., Abrudan, I., Barbati, A., Barreiro, S., Bielak, K., Bravo-Oviedo, A., Ferrari, B., Govedar, Z., Kulhavy, J., Lazdina, D., Metslaid, M., Mohren, F., Pereira, F., Peric, S., Rasztovits, E., Short, I., Spathelf, P., Sterba, H., Stojanovic, D., Valsta, L., Zlatanov, T., Ponette, O.: Knowledge gaps about mixed forests: What do European forest managers want to know and what answers can science provide? Forest Ecol. Manage. **407**, 106–115 (2018) (ISSN 0378-1127)
25. M. L. Lopez C.: 8th Forest Plan, Yamagata Field Research Center, Yamagata University University Forest, Watershed Preservation Section (2014)
26. Agisoft Company: (2016). Accessed 15 May 2017. www.agisoft.com/
27. Jégou, S., Drozdzal, M., Vázquez, D., Romero, A., Bengio Y.: The one hundred layers tiramisu: fully convolutional DenseNets for semantic segmentation. In: 2017 IEEE Conference on Computer Vision and Pattern Recognition Workshops (CVPR Workshops 2017), Honolulu, HI, USA, July 21–26, pp. 1175–1183 (2017)

28. Yu, F., Koltun, V.: Multi-scale context aggregation by dilated convolutions. In: 4th International Conference on Learning Representations, ICLR 2016, San Juan, Puerto Rico, May 2–4, 2016, Conference Track Proceedings (2016)
29. Chen, L.C., Papandreou, G., Kokkinos, I., Murphy, K., Yuille, A.L.: DeepLab: semantic image segmentation with deep convolutional nets, atrous convolution, and fully connected CRFs. IEEE Trans. Pattern Anal. Mach. Intell. **40**(4), 834–848 (2018)
30. Olafenwa, J., Olafenwa, M.: "FastNet." CoRR, vol. abs/1802.02186 (2018)
31. Ronneberger, O., Fischer, P., Brox, T.: "U-Net: convolutional networks for biomedical image segmentation. In: International Conference of Medical Image Computing and Computer-Assisted Intervention 18 (MICCAI), pp. 234–241 (2015)
32. Jarrett, K., Kavukcuoglu, K., Ranzato, M., LeCun, Y.: In: Proceedings of the IEEE International Conference on Computer Vision, pp. 2146–2153 (2009)
33. Srivastava, N., Hinton, G., Krizhevsky, A., Sutskever, I., Salakhutdinov, R.: Dropout: a simple way to prevent neural networks from overfitting. J. Mach. Learn. Res. (JMLR) **15**, 1929–1958 (2014)
34. Taniguchi, T., Tamai, S., Yamanaka, N., Futai, K.: Inhibition of the regeneration of Japanese black pine (*Pinus thunbergii*) by black locust (*Robinia pseudoacacia*) in coastal sand dunes. J. For. Res **12**(5), 350–357 (2007)
35. Jung, S.C., Matsushita, N., Wu, B.Y., Kondo, N., Shiraishi, A., Hogetsu, T.: Reproduction of a *Robinia pseudoacacia* population in a coastal *Pinus thunbergii* windbreak along the Kujukuri-hama coast. Japan. J. For. Res. **14**(2), 101–110 (2009). https://doi.org/10.1007/s10310-008-0109-1
36. Tinya, F., Márialigeti, S., Bidló, A., Ódor, P.: Environmental drivers of the forest regeneration in temperate mixed forests. Forest Ecol. Manage. **433**, 720–728 (2019) ISSN 0378-1127.https://doi.org/10.1016/j.foreco.2018.11.051
37. Krasnova, A., Kukumägi, M., Mander, Ü., Torga, R., Krasnov, D., Noe, S.M., Ostonen, I., Püttsepp, Ü., Killian, H., Uri, V., Lõhmus, K., Sõber, J., Soosaar, K.: Carbon exchange in a hemiboreal mixed forest in relation to tree species composition. Agric. For Meteorol. **275**, 11–23 (2019) ISSN 0168-1923. https://doi.org/10.1016/j.agrformet.2019.05.007
38. Jiang, X., Huang, J.G., Cheng, J., Dawson, A., Stadt, K.J., Comeau, P.G., Chen, H.Y.H.: Interspecific variation in growth responses to tree size, competition and climate of western Canadian boreal mixed forests. Sci. Total Environ. **631632**, 1070–1078 (2018) ISSN 0048-9697. https://doi.org/10.1016/j.scitotenv.2018.03.099
39. Lisein, J., Michez, A., Claessens, H., Lejeune, P.: Discrimination of deciduous tree species from time series of unmanned aerial system imagery. PLoS ONE **10**(11), e0141006 (2015). https://doi.org/10.1371/journal.pone.0141006
40. Vítková, M., Tonika, J., Müllerová, J.: Black locust—Successful invader of a wide range of soil conditions. Sci. Total Environ. **505**, 315–328 (2015) ISSN 0048-9697. https://doi.org/10.1016/j.scitotenv.2014.09.104
41. Kamilaris, A., van den Brik, C., Karatsiolis, S.: Training deep learning models via synthetic data: application in unmanned aerial vehicles. In: The Workshop on Deep-learning based computer vision for UAV, Proc. of CAIP 2019, Salerno, Italy, Sept (2019)
42. Kamilaris, A.: Simulating training data for deep learning models. in the machine learning in the environmental sciences workshop. In: Proceedings of EnviroInfo, Munich, Germany, Sept (2018)

Correction to: Developing a Configuration System for a Simulation Game in the Domain of Urban CO_2 Emissions Reduction

Sarah Zurmühle, João S. V. Gonçalves, Patrick Wäger, Andreas Gerber, and Lorenz M. Hilty

Correction to: Chapter "Developing a Configuration System for a Simulation Game in the Domain of Urban CO_2 Emissions Reduction" in: A. Kamilaris et al. (eds.), *Advances and New Trends in Environmental Informatics*, Progress in IS, https://doi.org/10.1007/978-3-030-61969-5_12

Chapter "Developing a Configuration System for a Simulation Game in the Domain of Urban CO_2 Emissions Reduction" was previously published non-open access. It has now been changed to open access under a CC BY 4.0 license and the copyright holder updated to 'The Author(s)'. The book has also been updated with this change.

The updated version of this chapter can be found at https://doi.org/10.1007/978-3-030-61969-5_12

A. Kamilaris et al. (eds.), *Advances and New Trends in Environmental Informatics*, Progress in IS, https://doi.org/10.1007/978-3-030-61969-5_19

Printed in the United States
by Baker & Taylor Publisher Services